P9-CDA-575

DATE DUE

BRODART, CO.   Cat. No. 23-221-003

DISCARD

813.54
H3458

Black Virgin
Mountain

Also by Larry Heinemann

Close Quarters

Paco's Story

Cooler by the Lake

Larry Heinemann

Doubleday
New York London Toronto Sydney Auckland

Black Virgin Mountain

A Return to Vietnam

PUBLISHED BY DOUBLEDAY
a division of Random House, Inc.

DOUBLEDAY and the portrayal of an anchor with a dolphin
are registered trademarks of Random House, Inc.

Excerpt from Le Duc Tho dispatch taken from *Giai Phong: The Fall and Liberation of Saigon* by Tiziano Tirziani, translated by John Shepley. Copyright © 1976 by John Shepley. Reprinted with the permission of St. Martin's Press

Book design by Terry Karydes
Map by Jeffrey L. Ward
Interior photographs by Franklin D. Rast

Library of Congress Cataloging-in-Publication Data
Heinemann, Larry.
Black Virgin Mountain : a return to Vietnam /
Larry Heinemann
p.   cm.
1. Heinemann, Larry. 2. Heinemann, Larry—Travel—
Vietnam. 3. Novelists, American—20th century—
Biography. 4. Vietnamese Conflict, 1961–1975—Veterans—
United States—Biography. 5. Vietnamese Conflict,
1961–1975—Literature and the conflict. 6. Vietnam—
Description and travel. 7. Americans—Vietnam. I. Title

PS3558.E4573Z48 2005
813'.54—dc22   2004061856

ISBN 0-385-51221-X
Copyright © 2005 by Larry Heinemann

All Rights Reserved

PRINTED IN THE UNITED STATES OF AMERICA

May 2005

First Edition

1   3   5   7   9   10   8   6   4   2

To Edie, as always,
and to Sarah Catherine and
Preston John. And to Asa Baber,
a brother of the blood.

Acknowledgments

Many thanks to the John Guggenheim Foundation and the William Fulbright Scholarship Program for their support. My heartfelt thanks to Kevin Bowen and the men and women of the William Joiner Center who have given me something I can never repay. My best and sincerest good wishes to Huu Thinh and the gracious men and women of the Vietnam Writers Association; to Dr. Le Van Thuyet, Director of the Office of International Relations of Hue University and to his superbly efficient and hospitable staff; to singer, composer, and spirited boon companion Vo Que of Hue City; to the kitchen folks of Vuon Lan Restaurant, and the cyclo guys and flower ladies of the Ben Ngu Market who treated me like family. One thousand blessings to you all.

— *Larry Heinemann*
Swift Lake, Wisconsin

Contents

There isn't a train I wouldn't take,
no matter where it's going.

— *Edna St. Vincent Millay*

We are not always right
about what we think will save us.

— *Bruce Weigl*

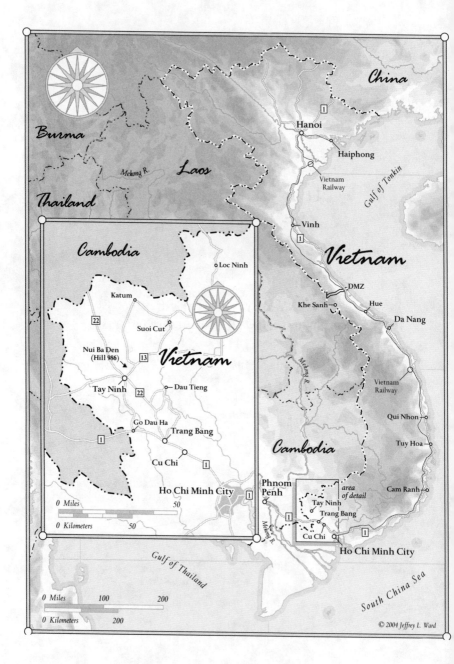

China

Burma

Hanoi

Haiphong

Mekong R.

Laos

Vietnam Railway

Gulf of Tonkin

Thailand

Vinh

Vietnam

Cambodia

Loc Ninh

DMZ

Khe Sanh

Hue

Da Nang

Katum

Suoi Cut

Nui Ba Den
(Hill 986)

Vietnam

Mekong R.

Vietnam Railway

Tay Ninh

Dau Tieng

Qui Nhon

Go Dau Ha

Trang Bang

Tuy Hoa

Cu Chi

Cam Ranh

Ho Chi Minh City

Phnom Penh

Cambodia

area of detail

Tay Ninh

Trang Bang

0   Miles   50

0   Kilometers   50

Cu Chi

Ho Chi Minh City

Gulf of Thailand

Mekong R.

South China Sea

0   Miles   100   200

0   Kilometers   200

© 2004 Jeffrey L. Ward

1

Several Facts

I was a soldier once, and did a year's combat tour in Vietnam with the 25th Infantry Division at Cu Chi and Dau Tieng from March 1967 until March 1968.

The town of Cu Chi, twenty miles or so northwest of Saigon, straddled Highway #1 (see map) and was profoundly undistinguished. The American base camp was just outside of town. Nowadays it is famous to the world for the Tunnels of Cu Chi, built by the South Vietnamese guerrillas with ordinary garden tools over a decade and more, and which spread out (if you stretched it) beneath us two hundred kilometers' worth. I am told that the local Vietnamese revolutionaries looked on in astonishment as our division engineers laid out and then built the base camp of considerable acreage over a portion of the

1

tunnels. This was not to be the last of the 25th Division's fuck-ups. Is it any wonder that when asked to describe the Americans during the war, about all that occurs to the Vietnamese is that we were "brave" and "valorous"? That's what armchair historians say about the Federal troops who assaulted the Stone Wall at the foot of Marye's Heights during the Battle of Fredericksburg in 1862, and who disappeared, said one participant, like snow falling on warm ground.

Dau Tieng was the base camp for the division's 3rd Brigade, squat in the middle of the Michelin Rubber plantation—forty miles north (and a touch west) of Saigon as the crow flies—in Cochin China; the classic image of a company town in every sense of the word. The Americans lived in run-down tents with dirt floors and slept on cots (the canvas all but rotting off the wooden frames), and shared the base camp with half a dozen large French colonial manor houses that had galleries all the way around where the plantation management and extremely senior brigade officers lived, tile-roofed plantation outbuildings, and an aboveground Olympic-size swimming pool (of all things); the lanes and gardens were lushly shaded with plane trees—just like in the movies. Outside the perimeter, the village streets were lined with offices, block-long clusters of company-owned housing, and somewhere in there was the ubiquitous company store. Down by the river was a huge latex processing plant that gave off a heavy industrial stink rivaled only by the leaden, acrid smell of foundries and mills in Southside Chicago and Gary, or the bourbon distilleries of Bardstown, Kentucky, on sour-mash day. The thick orchards of working rubber trees came nearly to the base camp perimeter, which was marked off with sloppy coils of concertina wire and spotted with sandbag bunkers, pathetic and well-weathered hovels that collected garbage and rats. The plantation ("the rubber," we called it) was laid out with cornfield-like precision that was seriously scary but somehow pleasing to look at; there was an undeniable parklike atmosphere. It should come as no surprise to hear that during the war the tending of the broad stands of rubber trees and the harvesting of raw latex diminished year by year, but it never ceased. War was war, to be sure, but business was (ever and always), of course, still business. Halfway through my tour we were told that the Army had to pay Michelin an indemnity for every rubber tree we knocked down—an easy thing to do with a thirteen-ton armored personnel carrier; a thousand dollars per tree, more or less. Well, after we heard that, we never missed a chance to take a whack at one. Fuck rubber trees; fuck the Michelin Rubber Company; fuck the Army.

In the spring of 1966 my younger brother Richard and I had received our draft notices, and we submitted to conscription with soul-deadening dread; Richard was twenty and I was twenty-two. No one told us we could hightail it to Canada. No one told us we could declare ourselves conscientious objectors and opt for alternative service—a special punishment all by itself during those years (like the preacher's son I know who did two years in a big-city hospital morgue; might as well have been Graves Registration). No one told us there were *any* alternatives. Even joining the National Guard, another well-known way of avoiding military service, was a waste of our time because everyone knew the waiting list was a mile long. You had to

be a well-connected politician's kid, some big-name professional athlete, or have some sort of clout otherwise. Such things
were not a topic of conversation in our family, anyway. Always,
the word in our house was: graduate high school; get a job. Ida
Terkel, Studs Terkel's wife, once asked what it would have taken
to keep my brother and me from going, and I told her that in
1966 she would have had to come into our house, sit down at
our dining room table, and explain it. All we knew was that if we
didn't show up for induction, a couple of guys from the FBI
would come looking for us, and off to jail we would go; and jail,
then as now, was no fun.

Our draft notices, literally facetious letters of congratula-
tion from President of the United States Lyndon Johnson, arrived in the same mail. Richard and I walked together through
our induction physical with one hundred other guys, passed together, took the oath together, were put on a train together (the
Illinois Central's City of New Orleans, as it happened), and
taken south for Basic Training at Fort Polk near Leesville,
Louisiana. Fort Polk, home of the Tiger Brigade, where the
11-Bravo light weapons infantry trained before going straight
overseas.* I was born and raised in Chicago, and I hadn't been
much farther away from home than St. Louis. Our family was
not much for traveling, and the farther south Richard and I
(and the rest of the conscripts) traveled, the more depressing
the countryside looked. Here was my first unsullied look at the

---

* Every Army job, called Military Occupation Specialty (MOS), had a number. The drill
was that everyone first went through Basic Training and then went on to his or her Ad-
vanced Individual Training—AIT. During the Vietnam War, the 11 prefix designated combat: 11-B (for Bravo) was light weapons infantry; 11-C, artillery; 11-D, reconnaissance;
11-E, tanker; etc. Infantry AIT at Fort Polk was called the Tiger Brigade.

rural, Southern poor; ramshackle farms with unpainted barns and swaybacked barbed-wire fences, dry-earth fields, and well-weathered farm machinery (the paint job all but burnt away). And it was hot; God, was it hot, and the rain came down in roaring sheets and filled the overlarge ditches to the brim. More than once the runoff came down the hill at the back of our barracks and washed in one door and out the other. Between downpours everything was dry and dusty, and crawling around the woods, training our little hearts out, everyone in the company agreed that Fort Polk was on the same list of shit-holes with Fort Leonard Wood in Missouri, Fort Ord in California, Fort Bragg in North Carolina, and Fort Hood in Texas. I looked around at the military squalor and thought to myself that when Louisiana seceded from the Union all those years ago, they should have declared New Orleans an open city and let the rest go. Richard and I were sent to the same training company, the same platoon, the same second-floor squad bay where we stood side by side in front of our bunks every morning for inspection. Our father, an awkward and uncommunicative man, sent self-conscious, not-quite-newsy letters; I would get the original and Richard the carbon copy. Our training company was made up of guys from Chicago and California. The draftees among us laughed loud and long at the Regular Army volunteers—the poor suckers who got conned into joining up; the Army was going to make a man of them; they were going to "learn a trade." That got a laugh every time. And, I kid you not, one of the California enlistees was a guy named Gump—"Like *gum* with a *pee*," he was always careful to explain.

Here we encountered what is perhaps the dumbest man I have *ever* met. Drill Sergeant S-----, one of those classic, dufus boneheads for which the Army is only too famous. It's the guys like him who wind up working as guards in military prisons— *that* he can handle. The kindest thing you could say about Drill Sergeant S----- was that he was not the sharpest knife in the drawer.

You need to know that during the post–World War Two Cold War years of Selective Service, every once in a blue moon twins would be drafted together, go through Basic Training together, but afterward split up and deliberately be sent their separate ways. This was because of the Sullivan Rule: in 1942 the five brothers named Sullivan of Iowa joined the Navy, and for sentimentality's sake were allowed to serve together aboard the cruiser USS *Juneau*; that November the *Juneau* was struck by a Japanese torpedo and sank along with seven hundred crewmen, including all five of the Sullivans—George, Francis, Madison, Joseph, and Albert (the youngest). The shock of that singular loss, not to mention the utter devastation to the family, caused the military to adopt the strict policy that blood kin brothers could train together but could not afterward be compelled to serve together—just in case—even in the same war zone.

Well, Richard and I stood side by side in formation and, aside from our white-cloth name tags over our right shirt-pockets, it was obvious by our brown hair, blue eyes, and the clefts in our chins that we were brothers. But I was two years older than Richard and half a head taller—we were definitely not twins. However, Drill Sergeant S----- simply could not get his mind around that fact; as if his imagination and view of the world would not permit it. He would eyeball us from under the wide, flat brim of his Yogi Bear drill sergeant's campaign hat. He'd look at Richard and his name tag; then he'd look at me and mine. Finally, S----- would say something like, "I know you two squirrels are trying to pull something, and when I catch you dipsticking around with whatever little scheme you've got cooked up, you are in big, big, big trouble."

What, I ask you, do you say to that? In later years we spoke of that clown many times.

Well, we horsed around Fort Polk for eight weeks, and about the only thing anybody ever got out of Basic was "in shape"; all those first-thing-in-the-morning miles of the Airborne shuffle, all those push-ups and jumping jacks and squat thrusts (PT, we called it); all those hours in the bayonet pit where we learned that the spirit of the bayonet was "to *kill!*"; all those sessions in the sand-filled hand-to-hand pit where we learned to "rip your head off and shit down your neck"; all those pugle-stick brawls; all those sessions at the rifle range; all of the brigade chaplain's character lectures. You learned to call everyone by his last name; even Richard and I called each other Heinemann, laughing. The fat guys thinned down and the skinny guys like me gained weight, and just about everybody wised up.

After Basic, Richard and I went our separate ways.

He was sent to the artillery at Fort Sill, Oklahoma, and eventually wound up in Germany on the crew of a Pershing missile. There is more to his story, and we will come to it by and by.

I was sent to Fort Knox (famous as the Gold Depository and "The Home of Armor") to join the armored cavalry—my specialty would be reconnaissance (Armored Intelligence; recon, we called it). Our cadre kept telling us that we were "the eyes and ears of the combat arms," trying to hide their smirks and instill in us skeptical draftees what passed for esprit de corps. A hundred-twenty-thirty-forty years ago, I would have been a scout in the horse cavalry. You know, the flinty Lee Marvin character in the John Wayne movies, looking sharkish in his mountain-man leathers with a beard like an inner tube, a juicy hunk of chew in his cheek, an eagle feather as long as your arm tied to his hair, and a hard-boiled voice straight out of a whiskey kegger. He'd be packing an Arkansas toothpick, a vintage model lever-action Winchester, a Henry buffalo rifle (one shot will bring down a twelve-hundred-pound buffalo), and a couple of six-shot 1851 Navy Colt revolvers in a carnival display of visceral panache. And he'd be mounted on some big-ass, steady-as-a-rock chestnut plow horse (called Bubbles or Hipshot or Standard Issue). After scouring the tall savanna grass of the Old-West Great Plains scaring up Indians (like a setter flushing pheasant), he'd ride back hell-for-leather and come pounding up to Mr. Wayne with hair a-flying and eyes wild, whooping and shouting, leaning way out of the saddle, "Balls o' fire, Kunnel! A whole *passel* of Injins! Lots of 'em. 'Ata way."

In Vietnam, to be called a "John Wayne" was a flat-out insult. Boyhood hero as verb: to *John Wayne* it, to *pull* a "John Wayne," was strictly for the ticket-punching lifers, the Boy Scouts, and the other assorted hot-dog, hero wannabes—the guys who had watched *way* too much television. You want somebody to take a mess of hand grenades and assault that bunker up the slope yonder? Well, sir, get John. He's just that big a fool, right down to pulling the pins with his teeth; dead already, went the running gag, but too dumb to lie down. John Wayne's real name was Marion Michael Morrison, and the man, the husband and father, the actor, was likely as friendly and likable and generous and gregarious as the day is long. But "John Wayne," the big-screen Technicolor postwar film persona and pop culture legend I grew up with, was something altogether different. His 1968 film *The Green Berets* is especially patronizing and insulting, though screamingly funny in a gallows-humor sort of way, right up to and including the closing image when he stands with his arm around a young Vietnamese lad at the very edge of what we are to suppose is the South China Sea, facing *east*, watching the sun "go down." In the late spring of 1979, I was driving home one night and happened to catch the *Chicago Sun-Times* headline in a street-corner vending machine out of the corner of my eye: JOHN WAYNE DIES. I started giggling, then laughing, then roaring with laughter. "John Wayne," the larger-than-life Hollywood character, the very beans of testosterone-poisoned, cartoon-macho movie bullshit, was dead; finally, and thank God. I laughed so hard that tears came to my eyes and I had to pull to the curb.

Military service has been important to the have-not, working-class young men in this country at least since the time of the famous Irish "famine boats," when the men were recruited into the Civil War Union Army right at the dock. Military service will always be seen as a "crucial rite of passage" for young men; if it was good enough for the old man and the uncles, it's good enough for me. Just finished with school, looking to get out of the house once and for all, ready to scratch the traveling itch and see the world? Might as well start with Germany. Ah, those lusty fräuleins; ah, the Bavarian Alps; ah, Oktoberfest. For some, military service is the only way up and out. And some are driven to military service by legal circumstances; it is understood as an all-but-fatal chore—just grit your teeth, cousin, gulp down your reluctance, and get through it. Regardless, here commences the rest of your education with a little something about how the world actually works—the banal, bland imagination and pervasive stupidity of large, severely organized hierarchies and the closely shaved, narrow-minded venal immaturity of the guys "in charge." And, if you don't already know, you learn that the world at large doesn't much care if you live a decent life or die in a ditch with a bullet in your head and dirt in your mouth. You are not so much a number, not even one of those famous bricks in the wall, as you are a cipher, like the tick of chalk on a dart game scoreboard. Our armed forces have always recruited heavily among the nation's high schools. Boards of education everywhere are only too glad to give the buffed-out, hardy-looking recruiters all the time they need for the shuck and jive pep talk, the four-color posters, and the slick commercial video showing the happy smiles of the Airborne Rangers as they shinny down ropes dangling into verdant pastures—and a lot of guys go for it. (Nowadays a lot of young women go for it, too.) Here is an endless supply of cheap labor that will work for peanuts, shut up, and do what it's told; all those kids "saving for college" the hard way, and no one paying the fine print any mind until it is long past too late.

At Fort Knox we trained with silly, pointlessly extravagant thoroughness for service in "Europe," undoubtedly in a NATO-garrison armored cavalry troop on the old Czechoslovakian border where, during our off-duty hours, we would carry on the fight against communism, swinging drunk, with our beer mugs; such were the expectations of Cold War draftees.

The officers at Fort Knox wore yellow cavalry scarves and thought they were either George Armstrong Custer or George S. Patton—the very poster boys of martial theatrics. The Gold Depository, the actual building, looked like a municipal cemetery mausoleum with a simple one-door loading dock, surrounded by many acres of well-kept lawn patrolled by large dogs and a high wrought iron fence; the actual gold, it was said, was stored in vaults deep underground, but in the seven months I spent at Knox I don't recall ever seeing a truck coming or going from the place. The roadside boulders around the fort were painted a dazzling sugar-cube white. The many, many yellowy cream—colored two-story barracks that lined the roads and lanes were the extreme holdovers of the "temporary" barracks built for World War Two—drafty, shacky dumps. Somewhere here is the barracks where Elvis Presley did his Basic Training, blessed with a plaque to commemorate that marvelous fact, and visited by the occasional carload of worshipful fans.

The crews of the tanks and armored personnel carriers prepared for motor pool inspections with top-to-bottom Simoniz wax jobs. The sharpest troopers among us put Johnson's Glo-Coat floor wax on our spit-shined shoes and boots, and stitched the creases into our inspection-ready, starch-stiff uniforms (which were often not worn at any other time), the trouser legs crackling when you stood on a chair to shove your feet and legs through. No one thought this absurd. These guys always looked *good*, were "squared away" and "outstanding," did every little thing by the very book, and then some. Or they were called simply "STRAC" (as in, "There goes one STRAC motherfucker"), an acronym the etymology of which is a complete mystery to me, but may well refer to what James Jones aptly called "a pointless pride." Take it all around, if the Army seemed a monumental waste of everyone's time (and the unadulterated, aggressively truculent chickenshit endless), you nevertheless did acquire a good deal of out-of-the-way knowledge that you weren't going to forget anytime soon (much like Mark Twain's story of the guy who once carried a cat home by the tail).

Our training finished, almost everyone was transferred overseas to Germany or Korea. Half a dozen of us were sent across the fort to D Troop of the 32nd Armor. The Fort Knox Officer Candidate School turned out bunch after bunch of crackerjack cavalrymen, called ninety-day wonders. The OCS guys used our tanks and tracks for training, and the 32nd was in the field a good deal. D Troop's barracks were across the street from a so-called remedial training company. These were the guys who simply could not hack it during regular Basic Training. It was painful to watch, and some of them could not even get the hang of so simple a thing as standing at attention, and otherwise seemed severely unsuited for military life. I have often wondered since if they weren't part of Secretary of Defense Robert McNamara's famous "100,000," men who were drafted when the physical and psychological requirements for service were severely relaxed as the need for warm bodies in uniform became more and more urgent. (There were eventually to be more than 350,000 of these guys, and it would be telling to know how many of them wound up in Vietnam and how many of them made it through in one piece, but no one paid much attention to those numbers.)

And down the street and around the corner was the WAC's, the enlisted women's, barracks. These women worked as eyecandy clerks and all around gofers down at the administration building and elsewhere around the fort. Every morning, except Sunday, everyone on the fort would fall out for reveille, and because of the women, none of us ever missed that first formation and first laugh of the day. We would listen as the WAC sergeant major, a large, thick woman, walked to the front of her formation and said in a large, operatic alto voice with plenty of coloratura oomph, "When I call attention, I want to hear four hundred pussies go *woof*."

We spent our weekends in Louisville, about forty miles north. Two or three guys would split a room at the famous old Brown Hotel on Fourth Street. The wide lobby was lined with equestrian portraits of Churchill Downs and the Kentucky Derby's more famous winners, standing just so. Saturdays the hotel was packed with fresh-scrubbed GIs in ill-fitting dress khakis, checking in and hanging out—me among them. The hookers were everywhere, and the strip joints and bar-fight saloons were clustered together down along the river. I stopped going to the riverfront strip clubs after the night I sat through the sad, creepy bump and grind of a young, five-months-pregnant black woman who gave us a detailed and graphic gynecological lecture-demonstration.

In the early spring of 1968 when I first returned home, I

headed straight for Louisville where my girlfriend, Edie, soon to be my wife, would meet me.

We had met Labor Day Sunday, 1966, in Bardstown, Kentucky, where she was a junior at Nazareth College, a Catholic girls' college in the same neck of the woods as the Abbey of Gethsemani where Trappist monk and writer Thomas Merton lived and worked. Bardstown is famous for its bourbon distilleries and the classic Southern antebellum mansion where Stephen Foster wrote "My Old Kentucky Home." Fort Knox is west of town about forty miles or so, and at night you could hear the soft, rumbling booms from the tank gunnery range, just yonder. The war had a strange effect on the school. A good many soldiers dated women from the college. One by one the boyfriends, and me among them, were sent to Vietnam, and every once in a while one of the women would get a letter back marked *DECEASED* across the address. More than one soldier never came back. Edie and I dated all that fall and into the winter. In February, when I was home on leave before going overseas, she drove to Chicago through one of those big-blow, late-winter blizzards that all but closed down the 270 miles of Interstate 65 from Louisville to Gary to spend a couple days with me. While I was overseas Edie wrote practically every day, and just before I left Vietnam I sent her some money to rent me a place in town.

I got back to my folks and my brothers and the house where I grew up on a Monday night, and all that week I slept on a blanket on the floor because I simply was not comfortable in my old bed. Plainly said, I couldn't get out of there fast enough, because it was clear that I would always be a boy in that house. Edie and I quickly arranged to meet on Saturday in the lobby of the Brown Hotel, the only place in Louisville I knew I could find, and, of course, on a Saturday afternoon the place was packed to the doors with weekend GIs. She and I sat way off to the side on a sofa, away from the hubbub, talking with all the anticipation and excitement that a year's absence naturally generates. What a meeting. A short while later the manager, circulating through the lobby, no doubt making a rough head count of the weekend's take, walked up to Edie and very sternly said, "How many times do I have to tell you broads not to work the lobby. Now, get out of here." Oh? I got right up and was within an ace of beating that smirking shit-ass half to death—"You fat-boy, Jody-fuck little weasel, who are you calling . . . ?"—when Edie grabbed my arm and got me out of there.

One of the very first ironies of the war: for someone who had never handled a firearm in his life, I became an excellent shot with the Colt M-1911 .45-caliber semiautomatic pistol (first used during the Philippine Insurrection), the M-14 assault rifle (which replaced the M-1 Garand of World War Two and Korea), the M-79 grenade launcher (basically a 40mm sawed-off shotgun), the M-60 machine gun (relatively lightweight and almost fuckup-proof), and the .50-caliber M-2 machine gun (the ones mounted on the World War Two bombers and big tanks). The "fifty," as we called it, was utterly dependable, never jammed or hung fire, threw a slug about the size of your thumb, and could just about blow your head off at any range; Joseph Heller's Snowden was a fifty gunner.

I could also read a map and use a compass (a considerable

skill if I do say so myself, and you'd be surprised how many people couldn't); could call in and adjust artillery; could work a 6omm mortar; could, in a tight spot, do medic's work; and could, of all things, use a Geiger counter to check for radioactive atmosphere in the aftermath of a tactical nuclear catastrophe should anyone be left alive enough to care.

I could type 40 wpm, drive a jeep and a deuce-and-a-half (about the size and heft of a three-yard dump truck), and spit-shine my boots and shoes within an inch of their lives. I could duckwalk into the horizon, and dismounted-drill around the parade ground screaming raunchy marching ditties at the top of my lungs (the Army's contribution to world literature) until I was blue in the face and my shins cracked, then stand *tall* with my heels together and eyes *front*, and whip the dictionary definition of a hand salute on anything that moved. Above all, I learned to beg for my pay, a thing my father, an ordinary working stiff, told me never, never, *never* to do. ("Payday's payday, son, they *owe* you the money!") We would line up alphabetically in our spiffiest dress uniforms, spit-shines, fresh haircuts and all, and one by one stand in front of the captain, whip a salute on the guy, saying "So-and-so reporting for pay, *sir!*"—everyone sounding off like he had a pair. The captain would eyeball you up and down, and if he didn't like the way your uniform looked, or if he just didn't like you, back you went to the end of the line. Otherwise, he counted out your pay in cash and had you sign the receipt. Then you went to the next table and kicked back $17.50 for your monthly U.S. Savings Bond that everyone was required to buy, anted up fivers for the Mess Hall Fund, the Day Room Fund, and so on.

On the other hand, I was not overfond of hand grenades;

you couldn't throw them far enough away to suit me. And, I am proud to say, I could *not* night march through the Okeedokee Swamp, nor kill an alligator with my bare hands, skin it with my thumbnail, and eat it raw, nor jump out of a perfectly good airplane screaming my goddamn fool head off, nor swim from Los Angeles to Catalina Island on one gulp (straight up to the beach with my K-bar combat shiv in my teeth), good to go.

Meanwhile, of course, there was a war in Southeast Asia—a spot of bother, as the good soldiers among the Other Ranks of the British Empire might have said—as meaningless and vicious a war as this country ever fought; a very good reason for our conspicuous lack of enthusiasm.

Just after the first of the year in 1967, my orders came down for Vietnam, and in early February I was given a month's leave before I had to report to Oakland Army Terminal "for transport overseas." The day I left Fort Knox our first sergeant called me into his office. First Sergeant Alva was a full-blood Navajo Indian, built like a fireplug with buzz-cut hair, who spoke with a thick, virtually unintelligible accent (it was definitely not Spanish)—one of the very few lifers I encountered in the Army who seemed to know what he was doing. He was gruff, proficient, and businesslike in the way that later reminded me of First Sergeant Warden in James Jones' *From Here to Eternity*. When I gathered my gear and walked into the orderly room to sign out, First Sergeant Alva invited me to take a chair in his office, the only time I could recall sitting in his presence. We talked about a good many things. Finally it was time for me to leave. We shook hands and he wished me luck, saying, "Remember, Heinemann, this is not a white man's war."

*   *   *

I wound up in the reconnaissance platoon of a mechanized infantry battalion (poor cousin to the regular armored cavalry;
that is, no tanks) at Cu Chi and then Dau Tieng. We rode M-113
armored personnel carriers, APCs—tracks, we called them; the
Vietnamese called them "green dragons."

In recon, four men rode on each of the ten platoon tracks.
The driver sat inside from the neck down on the left next to the
engine, the track commander stood in the armor tub behind
the fifty mounted on the main hatch, and two observer-
gunners sat between the smaller M-60 machine guns (pigs, we
called them) toward the back; a configuration called an ACAV
(armored combat assault vehicle). We might as well have been
mounted on rolling bunkers. Recon seldom did much actual
reconnaissance that I recall, except what cavalrymen refer to
with sly irony as reconnaissance-in-force. We escorted truck
convoys from Dau Tieng to Tay Ninh and back; one round trip
was a day's work. When the battalion went to the field, we went
along to set up a perimeter, called a laager, at a forward support
base or LZ (landing zone) for the 105mm or 155mm self-
propelled howitzer batteries that provided close-in artillery
support for the line companies—the outgoing salvos screaming
just over our heads. Or we worked directly with one of the line
companies in free-fire zone sweeps and search & destroy missions, but just as often we worked by ourselves. And rarely, as I
say, were we given an actual reconnaissance mission.

In the four line companies, the crew of a track consisted of
a driver, a track commander on the fifty, and a squad of ten in-
fantrymen. Four tracks made up a platoon, four platoons made

a company, and four companies made a battalion. As well, in the battalion there were the medic tracks, the 81mm mortar tracks, the flamethrower tracks (which could throw a stream of flaming napalm quite a distance), the 106mm recoilless rifle jeeps, the tracks for the company commanders and the colonel and his staff, and so on. On the road the battalion made quite a parade, stretching perhaps a mile and more.

Simply put, our tracks were death traps; boxy, squat, and ugly thirteen-ton contraptions put together with flat slabs of inch-and-a-quarter aircraft-quality aluminum alloy armor plate, off-the-shelf Chrysler V-8 engines, blowers on the four-barrel carburetors the size of ceiling fans, straight pipes that came right off the exhaust manifolds, eighty-gallon gasoline tanks, and these wanky little two-dollar horns. We never surprised anybody. The tracks would only do thirty miles an hour top-end, but we drove them full-bore flat-out every chance we got. It was a great kick to take these large, loud machines into town, park anywhere we goddamned pleased, hit the cafés and whorehouses, and generally take a break in place—as we called it.* Neither the MPs nor the Vietnamese cops bothered us. These machines were obsolete the day they were built, though you see redesigned diesel-powered models in the war news all the time.

Each track carried thousands of rounds of machine gun ammunition; enough to last all day. When you fired the fifty, the hard, blunt shiver of the recoil went straight up your arms, all but blurred your vision, and your dog tags literally danced and jangled on your chest. To get a sense of what it was like to

---

* *I've spoken to a number of Gulf War veterans and they said that it was, indeed, an equally great kick to drive their M-2 Bradley Fighting Vehicles (very similar to our tracks, if larger and even more heavily armed) flat-out across the desert.*

work a fifty, watch and listen to the street construction guy operating the jackhammer as he rips up pavement. When all ten
recon tracks came on line for a mounted assault and all thirty
machine guns cranked off a sustained, rolling volley as the
tracks moved slowly forward, the air simply crackled and
pounded with noise and muzzle-flash flames; it gave you the
sense that he who makes the most noise, wins. Night assaults,
though rare, were especially horrific. Besides the machine gun
ammunition, each track carried several cases of hand
grenades, perhaps a hundred pounds and more of C-4 plastic
explosive (with a small wooden box of blasting caps), dozens of
Claymore antipersonnel mines, and hefty spools of detonation
cord, which for all the world resembled ordinary clothesline;
we rolled the C-4 into golfball-sized hunks to boil our water for
coffee. For sidearms we had M-16 assault rifles (universally regarded as junk), 12-gauge pump-action shotguns (with bagsful
of over-the-counter, double-ought buckshot shells—just the
thing for ambush), M-79 grenade launchers, .45-caliber pistols, and other personal weapons brought from home (small-
bore revolvers or serious knives). And floating around were
copies of the good old M3-A1 (called grease guns in the movies,
.45-caliber semiautomatics with thirty-round magazines), and
Chinese-made AK-47s. These were knockoffs of the old Soviet
design; the AK has been the issue assault rifle in the Soviet,
now Russian, army since 1949. To this day the automatic
Kalashnikov is the best infantry assault rifle in the world, the
choice of many a revolutionary army (watch the news). We
would pick up our AKs from Vietnamese "body count" after a
firefight.

    During those years of foreign war and home-front social

turmoil, the most famous photograph of an American with a Kalashnikov was taken during the 1973 American Indian Movement "occupation" of Wounded Knee on the Pine Ridge Reservation in South Dakota; group shot—one of the AIM long-hairs holding the AK skyward, head thrown back, obviously giving a very satisfying shout. The flat-out irony is, of course, that in 1973 the only way for an American to get himself an AK was to shoot somebody and take his.

Aside from the AK-47, the Viet Cong had an ad hoc variety of sidearms and rifles (American, acquired "by art," as well as French and Chinese), including the shoulder-fired RPG (armor-piercing rocket-propelled grenade)—something like the bazooka of World War Two, only smaller and more handy, an antitank weapon of singularly blunt and functional ugliness. You see RPG launchers on the war news a good deal, always carried with haughty and casual disregard. The RPGs were used against our tracks and they were "very effective," to use that arcane phrase of the time. Basically, an RPG would go through our inch-and-a-quarter armor plate like spit through a screen, and if the round hit the gas tank the track would go up like the head of a kitchen match, greasy mushroom cloud and all (the shit would *fly*); and, invariably, the drivers were incinerated on the spot. (An RPG will pretty much do the same thing to the modern twenty-five-ton M-2 Bradley Fighting Vehicle, which is constructed of the same inch-and-a-quarter armor plate.) One gallon of gasoline is equivalent to nineteen pounds of TNT, mind you, and will lift one thousand pounds one thousand feet into the air, instantly. Death traps, as I say. (In Iraq, RPGs have brought down more than one helicopter, wrecked more than one track and Bradley, totaled more than one Humvee; a nasty weapon, both dirt cheap and very portable, and it doesn't take a genius to fire—just load it, cousin, point and shoot.)

I recently met a man I hadn't seen since the day I walked away from the platoon in March 1968. One afternoon weeks before, his track had taken an RPG, and the thing had missed taking his leg off by *that* much, but whacked him with a hefty handful of shrapnel. For years after, he told me, he carried around a pair of tweezers, and every once in a while a sliver of shrapnel would break the skin and snag his trouser; he'd grab his pants leg, limp around until he found a place to sit, take out those tweezers, pull up his pant leg, and whip that sucker out.

Nevertheless, we were "mounted," felt virtually invincible and untouchable, and literally looked down on everyone. We did not envy the straight-leg, ground-pounding infantry at all, who, it seemed, kept getting into trouble and were dropping like flies. I kept the thought during my whole war-year that my track was my ride home; to my mind absolutely anything beat walking.

By the way, we called ourselves troopers, never grunts. "Grunt" was a slang word used to describe the Marines, up-country, and didn't become common usage until after I'd been home a couple of years. But, like the "doggies" or "dogfaces" of World War Two, we were treated like meat and expected to behave like meat, so "grunt" still sums it up nicely.

We never referred to each other as "buddies," that irksome and denigrating journalists' cliché. "Buddy-up" was something you did at Boy Scout camp when it was time to go down to the lake for a swim. Vietnam during the war was many things, but it certainly was not the Annual Jamboree (though there were plenty of Boy Scouts around, and the merit badges were handed out like popcorn).

Everybody kept count of the days left in his tour with one kind of calendar or another, some pornographic or downright obscene, penciling in one numbered square at a time until nothing was left but the bushy groin, and you packed up, threw the calendar away, and went home. Some used a regular pocket calendar and drew a large X on each date. I kept count out loud, saying the number each morning, "Only (so many) left to go." The 300s were always good for a laugh, and the 200s were one long pull. Once you hit 99 days, you were called a "two-digit midget," a short-timer, or simply "short," though by strict custom you were not "the" short-timer until you were next in line—even if it was only for the day. (More than one Stateside outfit had a short-timer's swagger stick, passed from one guy to the next.)

And we smoked marijuana, you may be sure—what we might call postmodern Dutch courage. Marijuana has a long and honorable history with the American military (as with other drugs, legal or not), was easier to obtain than PX beer, and you didn't have to keep it cold (not that the beer was ever cold, either); trust me, I had such a straight-arrow upbringing that I never *heard* the word marijuana until I went overseas. We smoked grass, "did" opium and hashish, dropped acid, and generally sucked down anything we could get the top off of; and *that* became more purposeful and casual, not to say routine, the deeper into your tour you got. If it seems reckless, not to say foolhardy and self-destructive, to be drunk or stoned in a war zone—particularly during a firefight (thoroughly bizarre occasions, those)—let us remember that soldiers have always been

drinkers; the stupefying numbness of a good "drunk" being
only too welcome. During the Civil War the Union troops drank
homemade brew called "Nockum Stiff," "Popskull," and "Oh!
Be Joyful." In the British trenches of World War One it was
common practice in many outfits to issue the troops a daily ra-
tion of strong, dark rum, which poet David Jones referred to
ever after as "the precious"; years later one man still remembered the trenches after a hopelessly bungled assault smelling
powerfully of rum and blood. James Jones spoke of the alco-
holics among the "old army," prewar lifers of Schofield Barracks swilling Aqua Velva aftershave (which tastes as god-awful
as it smells), and later on in Guadalcanal of a horrible fer-
mented concoction called "Swipe," made from #10 mess-hall
cans of fruit. In Southeast Asia, the village kids would bring
beer and marijuana to us in the field—everyone knew where we
traveled and camped—and it was not uncommon to find little
plastic bags of marijuana on the Vietnamese body-count
corpses, along with wallets and money and such. Hanoi writer
Bao Ninh,* the author of *The Sorrow of War* who served in the
People's Army of Vietnam (which we called the North Vietnamese Army, the NVA) from 1968 to 1975, wrote of soldiers
smoking the dried, shredded root of a mildly intoxicating
cannabis plant that grew wild in the woods, and he told me once
that even standing in the midst of a field of its blossoms in-
spired thoughts "of love."

* *The Vietnamese express the family name first and the given name after, the reverse of
Western custom. It takes a little getting used to, but here and after I will give Vietnamese
names in the Vietnamese fashion. Bao Ninh was an ordinary soldier like myself, and I
once asked him what was the hardest thing he ever had to do during the war, and he
simply said, "Bury all of my friends."*

I arrived in Vietnam scared to death, and, to make a long story short, we were not pleasant people (down where the rubber met the road, so to speak) and the war was not a pleasant business; what happened there is not pleasant to recall. We generally rode roughshod over the countryside around Cu Chi and Dau Tieng, the Iron Triangle and the Ho Bo and Bo Loi Woods, Tay Ninh City and the Black Virgin Mountain, and I have no doubt we radicalized more southern Vietnamese to Ho Chi Minh's nationalist revolution than we "saved." We understood perfectly well that we were the unwilling doing the unnecessary for the ungrateful. There is an old saying that sometimes the best way to disobey an order is to obey it to the very letter, and what Michael Herr said in *Dispatches* was true: we wanted to give Johnson and McNamara and the rest of that bunch a Vietnam they could put in their ashtrays.

We did not have what anyone could remotely regard as ordinary human contact with the Vietnamese, just as "Vietnam" was distinctly not a country, but, rather, an event; a war of the death's breath nightmare kind. We were severely and earnestly warned *not* to "fraternize" with the Vietnamese; not to learn their language; not to eat their food; not to listen to their music, or in any way come to appreciate them as a people or a culture, which goes back to the establishment of Red River Delta wet-rice farming toward the tail end of the Neolithic Age, you understand. According to the traditional story, the Hung kings of Tonkin established Vietnam's first imperial rule about 4,800 years ago; that dynasty lasted from 2879 to 258 B.C.E. (when the Vietnamese gave way to the Chinese, at once welcomed and despised); not a bad run, as these things go.

By the end of my tour in March 1968 (a month and a half

into the Tet Offensive, which for us started around Thanksgiving 1967 and lasted well into the summer), I didn't know anything more about Vietnamese language, military or social history, literature or music, religion or customs, or "sociology" than when I arrived.

Or about Cambodia or Laos, for that matter.

Tet, the Vietnamese New Year according to the lunar calendar, fell on January 31 that year, and the campaign that came to be called the Tet Offensive marked the simultaneous attacks against South Vietnamese cities from Hue to the city of Ca Mau in the extreme southernmost part of the Mekong Delta, and scores of places in between. But there had been hard fighting from one end of the country since long before Christmas, and the battles both large and small dragged on for months.

By the end of my tour in early March, I say, what I *had* learned was that the Vietnamese made great beer, Ba Muoi Ba (Vietnamese for 33, pronounced by us "ba me ba"), served in brown, one-liter bottles; that tailor-made, opium-soaked marijuana cigarettes (OJs, we called them) could be had somehow perfectly resealed in their brand-name cellophane packs; and that a countryside hooker would cost you a five-dollar bill (short-time, we called it). The young women and their snazzy, Saigon-cowboyish pimps followed our platoon around on Honda scooters and waved us into the hedgerow bushes with hearty, welcome smiles. And the everyday, garden-variety straight-vanilla workaday rule was that if anything human turned up dead, body-count it, and do not ever, *ever* leave any live wires behind. The permissions for what can be called "misbehavior" were broad. Leave it at this: we stood in the very midst of an out-and-out spirit of atrocity (which pervades every war, mind you), and we drew it down, down, deeply down into our lungs with every breath of that heavy, earthy Southeast Asian air.

And truth be told, I did not *want* to know about the Vietnamese, much less understand; still less appreciate. Even so, it seemed clear to us that the war was not simply a pointless waste, but egregious and iniquitous. Though those were not the words we used. ". . . the fuck outa here, bub. Fuck this." No one I knew wanted any part of it, and if there was any antiwar feeling among us, it didn't get much more sophisticated than, *"This is* bull*shit!"* In June 1967, we read with intense appreciation and no small jealousy that the Israeli Army had fought and finished a war against the Arab nations in six days (nice work, guys). And that fall we heard that a huge antiwar protest rally in Washington, D.C., had gathered round the Pentagon to levitate that building three hundred feet into the air and so exorcise the very demons within; great gag, that. Ed Sanders and the Fugs, couples fucking in the grass, chants of "Out, demons, out!," wild beatings and mace, and a bunch of troops from the 82nd Airborne ("Death from above") waiting inside the building ready, willing if not downright eager to get their hands around the throat of any poor fool who might actually make it through the door and shit-stomp the living daylights out of him. We read that the demonstrators had written War Sucks and Pentagon Sucks in large and hasty, graffiti-style letters and had pissed en masse on the stonework—"Ready, *whiz!*" We imagined the veritable rivers of pee washing over the building's façade, like the rivers-of-pee stories we heard from the fucking new guys just in from Oktoberfest. Well, we laughed so hard we squirted beer up our noses.

Like everyone else I simply wanted out, and when the end of my tour finally arrived, I turned my back and left that place with the wistful notion that when I got home I could act as if my war-year was as unremarkable and unreal as if it had happened to someone else; as if the war and my part in it had *never* happened; as if the whole mad, ugly business had been a dream, and I could pick up my life where I had left it. That didn't happen, of course (for a rifle-soldier no such thing is possible), but that was the naïve impulse that got me aboard my homebound plane at Tan Son Nhut.

Oh, and by the way, it bears mentioning that when our plane finally arrived at Travis Air Force Base and we were bused to Oakland Army Terminal where we would muster out, no one among us was hankering for a parade; didn't want a parade; didn't need a parade; fuck a parade, just one more goddamned formation. It was all we could do to sit still for the horsy nonsense of one last chaplain's lecture. I'll take my Discharge right here and now, *pal*, thank you very much, whatever pay you owe me, sign your little book one last time, and be on my way, but you can stick your parade up your ass, colonel or general or whoever the fuck you are. Years later—Father's Day Weekend 1986, as a matter of fact—there was a "Vietnam Veterans Welcome Home" parade here in Chicago, but I *still* could not bring myself to "march"; besides, the one and only William Westmoreland was to be the parade marshal, and I wasn't going to get in line behind that man if he were only going to the shithouse. Still, it was a remarkable party, perhaps the only time in my life I could recall thousands of grown men walking around downtown swinging cold six-packs in their hands. A friend of mine walked the reverse route of the parade, and every time he met somebody he knew, they'd drop out, find a saloon, and crank a few; I bumped into an ex-Marine on a Harley who had ridden all night from Downstate, hit town just in time to swing his bike into the parade, and idled and crackled his way along for the several hours it took the parade to end. I saw a woman on LaSalle Street young enough to be my daughter time and time again jump off the curb and walk backward in front of someone who caught her attention, crouch some, and take his picture with her big old Polaroid; then walk with him, talking, waving the picture in the air while it developed, hand it to him, whip a quick hug and kiss on the guy, and then hop back onto the curb and melt again into the crowd— slowly but surely working her way through the parade one photograph at a time.

For me, that whole weekend was a celebration rather than a welcoming home; after all my war-year was almost twenty years past. By 1986, I didn't need to be welcomed home with a parade any more than I did in 1968. The actual parade seemed gratuitous, and sad; standing at the curb I could not bring myself to cheer, any more than I could bring myself to join the cheer that went up in the plane that brought me home as it lifted clear of Tan Son Nhut runway (and we steeply climbed to altitude). But it was a personal occasion, nevertheless, and worth celebrating. This was, after all, Father's Day weekend. Here were my wife and kids. Here were good friends of mine, the best of my friends, in town to party. I had that Wednesday sent the manuscript of my second novel, *Paco's Story*, to my editor, and whatever the book means to anyone else, I knew that I had turned in a good piece of work; it was virtually a physical satisfaction. Isn't life a good, sweet thing?

*   *   *

It is a considerable understatement to say that the late 1960s and early 1970s was a time of great social change in this country; not to say upheaval. Cultural catastrophe of one kind or another accompanies every era of war, and our war in Vietnam was no different. The world seemed to be coming apart; even the most deadened, flat-lined imagination understood that *some*one's hash was going to be settled, and that right soon. (Think of the Prague Spring and the ruthless response of the Soviets to the Czechoslovakian push for social reform; think of the Paris revolution, called the merry month of May, and those bitter street battles.) Human history convulsed once again, and the dead-hard crack of it reverberated out the same as lightning slams into rock—chips and chunks will fly (*and* sparks), and you feel the dumb shock of it in your feet, up through the knees to the very eyeballs and the roots of your hair (and the cemeteries fill to bursting as far as the eye can see). What with the assassination of Martin Luther King Jr. not three weeks after I got home (and the immediate aftermath of self-destructive, enraged grief that wasted many a black neighborhood in many an American city) and then in June the murder of Robert Kennedy (shot in the back of the head like his older brother, President John Kennedy), and the increasingly strident and sharply antagonistic relationship of the government with the civil rights movement and the antiwar/peace movement—with all that, I say, by the end of summer 1968, it seemed as if the war had followed me home, and I wanted nothing to do with it.

Just out of the Army and newly married, I got a job that summer driving a bus for the Chicago Transit Authority (the

CTA)—the sort of work my father did all his life. The summer-job college guys picked up the slack for the regular drivers on vacation. Most strangely, that summer, there was a group of Iranian foreign exchange students, all petroleum-engineering majors from the University of Missouri School of Mines in Rolla. Thank God the CTA gave out route maps and very specific directions, because those guys had no idea where they were. Several had never driven a car before. One guy spoke such poor English that he wore a button on the side of his hat that read, "I am a deaf mute." The money was absolutely righteous—I made more money that summer than I did in two years in the Army—but I was your worst bus-driver nightmare come true; I drove my bus like a bulldozer. I came to appreciate the source of my father's fierce and violent temper; somewhere here was the bedrock of all those belt-whippings my brothers and I endured when we were kids; and that summer I was struck, suddenly, by how somber and small he seemed to me. In August, during the now-famous Democratic National Convention, I drove through the fringes of the big doings in Lincoln Park when the cops "cleared the park" at 11:00 P.M. (enforcing what was virtually new law; the night air chilly as always, and now peppery in the eyes and nostrils with tear gas; CS, we called it).* Later in the week I rolled through the extreme edge of the confrontation between the demonstrators and the cops at the Conrad Hilton

---

* CS is the abbreviation for O-chlorobenzyledide malonontrite. Developed in the 1950s, "CS won't kill you" goes the story, but it will make you very sick. We used CS grenades to "smoke" tunnels rather than just toss a couple frags. Then in the fall of 1967, protests of the use of chemical warfare persuaded the Pentagon to order the halt to using CS, so we used the gas we had and afterward switched to straight frags for tunnel work. It was not long after that urban police forces discovered the beauty of CS to deal with "street work," and tear gas is still in use today.

and was flat-out chilled and appalled. The cops-on-kids "massacre" conveyed the image of an extemporaneous berserk mêlée, but the cops had been repeating all that week that they were just "following orders" to take off after anybody who got in their way as they shoved people around with their nightsticks, or cleared Lincoln Park, or just plain stomped *any*one they pleased—kids and journalists and Lincoln Park neighborhood folks alike (even Hugh Hefner got whacked on the ass).

The night Hubert Humphrey got the nomination, many, many demonstrators gathered at Michigan and Balbo in front of the Conrad Hilton Hotel where most of the convention delegates were staying. Several blocks away, I hit the corner of Clark Street and Congress Parkway; a wide boulevard with a generous center curb. And there, parked bumper-to-bumper along the five blocks of curb from the LaSalle Street Station to the "L" structure at Wabash, were many, many buses with engines shut down and lights off (a big no-no for drivers), each filled with cops (swinging loads, bus drivers call them). Perhaps a thousand cops and more were tricked out in riot gear, the uniform we'd been seeing all that week—all summer, in fact—some standing by platoon on the sidewalk, some sitting in the buses. I sat at the light, leaning over the wheel (elbows and hands), and whatever pristine certainty I possessed—shucked of *all* doubt—about the war as a great wrong thing and the rightness of anyone's opposition to it (political, moral, abject laziness, or outright cowardice); what*ever* was that extreme focus of clarity, I took one look at the cops arrayed before me in their hurry-up-and-wait, bored-out-of-their-fucking-skulls exasperated rage and knew exactly what was coming. The cops, as a bunch, looked like they couldn't wait to lay into those kids (stick-time, MPs called it).

I had *seen* this, done *this*, a time or two: the whole battalion (four line companies, recon and mortars, recoilless rifles and flamethrowers, the medics, the colonel himself, staff and all) would gather in those dark and quiet hours of a morning, then at the first hint of light encircle some godforsaken village out in the middle of nowhere with an inwardly directed cordon sanitaire, pounce on it, and spend the rest of the day conducting the war's work—right down to the dirt.

That night of cops and kids, the National Guard and network television, the whole world watched what became known ever after as "The 1968 Democratic Convention Police Riot." I knew exactly what was coming, I say, and I did not want any part of it; no thank you. When the light changed I tore out of there, swung a wide left, blew the light at Dearborn Street, hung *that* left, drove north past Van Buren Street, Jackson Boulevard, Adams and Monroe, Madison and Washington, Randolph and Lake, and kept going (blowing lights and breezing stops) until I was through the Loop, past Wacker Drive, and over the river. I don't know what *that's* called nowadays, but when I was growing up we called it hauling ass.

And as those first months and then years back from the war passed, regardless of my own hard-earned opposition to the war, I simply could not bring myself to join the Vietnam Veterans Against the War (even so little a thing as that and as much as I agreed with them); the VVAW—nonviolent antiwar ex-GIs trying to educate the home folks the best way they knew how about the true nature of the war. I did not travel to Detroit to listen to and watch the VVAW's "Winter Soldier" atrocity hearings in early 1971; nor did I show up in Washington, D.C., that April for Dewey Canyon III where, among other things, the veterans roundly denounced the war and threw away their medals* (the essence of their message was precise: mean things have been done, from the Oval Office on down, and in the spirit of meanness); nor a month later did I march with the VVAW from Concord to Lexington and Bunker Hill (precisely reverse of Paul Revere's "ride," holy ground of our own Revolution)—a demonstration which, by the way, resulted in the largest mass arrest in four hundred years of Massachusetts' history. I know a couple of guys who did march, and they got their asses royally kicked. A good friend of mine once told me that he would have been there, but that he was just then in the stockade at Fort Devins for going AWOL. Regardless of the sincere, bone-deep empathy I had with the VVAW, it was my understanding that the organization was basically run by ex-officers, and I'd had enough of lifers to last me quite a while (no offense, guys).

During those years, in fact, I rarely stepped off the porch.

Returning American soldiers caught a "home front" public raised on the patriotic fairy stories and lingering nostalgic assumptions of World War Two by surprise, and I think people generally didn't know what to make of us; the cherished mythology of an immature and grandiose national egotism dies

---

* One of the most telling "confrontations" during the Washington demonstrations occurred between several women from the DAR and the VVAW marchers. One woman remarked that the antiwar cadence chant ("One, two, three, four/We don't want your fuckin' war") was "obscene." Another woman scolded a marcher by saying, ". . . what you're doing is not good for the troops." To which the guy replied, "Lady, we are the troops."

hard. And whether it was real or not, we felt a chilly antipathy (to say the least) from the hardest heads in the antiwar movement; we felt looked-down-upon, at first, by the student activists; though I told the antiwar kids around me at Columbia College here in Chicago that they didn't know the fucking half of it. When I arrived back in the States no one spit on me or said anything to my face, but I was absolutely in no mood for such shenanigans—you may be sure. The Spanish news is that spitting on someone and shooting them with a rifle comes from the same place in the heart, and do not ever forget that. And too, I had just come from a place where I didn't take any shit from anybody. What possessed anyone in this neck of the woods to think that I was going to take any shit from him? If you had spit on me at the San Francisco airport or O'Hare, well, I would have kicked your ass from here to Chinatown—as the saying goes. That hard and shrunken hollowed-out look did not rub off quickly and genuinely took people aback. The thousand-meter stare (as we called it) is a bluntly intense, narrowly absorbed concentration; perfect in its all-pervasive, unambiguous vacancy without warmth or light. A most uncomfortable gaze, to put it mildly, when it settles on you; as many a writer has said many a time, a literal black look. And it was irresistible, you understand; by and by, you gave yourself up to it, and you even said it out loud, "I just don't fucking care."

And I don't know about anyone else just returned from overseas, but I felt joyless and old, physically and spiritually exhausted, mean and grateful and uncommonly sad; relieved as if a stone had been lifted from my heart and radicalized beyond my own severely thinned patience; pissed-off and ground down by a bottomless grief that I could not right then begin to express—"impacted grief" it has been called; a phrase that evokes that moment just before a boil as big as a nickel finally bursts in the mouth. If it is possible for a moment to speak metaphorically of the government as a "father"—who you can rightly assume loves you and wishes you only the best—well, to be betrayed by your father (not once, but three hundred sixty-five times) is a powerful unforgettable lesson, not unlike belt-whippings. I was a lab animal who had made it out of their "lab" in one piece, but what to make of all that bloody murder and heart-killing madness? (There was an awkward uneasiness in me that persisted for weeks because I wasn't wearing even so little a thing as my Filipino bowie knife; a very serious blade, that.) Coming home, I looked around and I knew for a fact that this was not my "place"; that I had not come home. Rather, I was deeply estranged, as if I had been taken out of time, knocked out, a disquieting, disorienting sensation. But home-bound soldiers have always felt thus. A writer friend of mine once told me that when her first husband came home from World War Two, he was fine for about a week, then laid himself on the couch, turned to the wall, and stayed that way "for about six months"; body-memory makes for long nights, and we have all heard the stories. After the Civil War, "it" was called soldier's heart; after World War One, shell shock and neurasthenia; after World War Two and Korea, combat fatigue—the boys are just tired. Nowadays it is referred to as post-traumatic stress disorder, but there is a taste of acerbic iron in the phrase, something peptic and sour, which simply does not fully convey "it." Poet and veteran Bruce Weigl has always referred to those several years just back from the war as "the black years."

So, trust me, wherever the world was headed in those years, it was going there without me; I had other matters to attend to.

On the other hand, I was not one of those guys who went to his room, started drinking his liver to death, and never said boo ("the war" eating him alive). I could not shut up about what I had seen, what I had done, and what I had become. I told anyone who asked, and without the polite dashes, the tasteful euphemisms, or the discreet ellipses.

The power to name is extraordinary, and during and after the war we soldiers were called many things. I wanted to name myself if you don't mind, and be precise; not a victim, certainly not a hero, just a man trying to deal with an uncommonly grotesque circumstance and come out the other end of it in one piece; however I had to do it and whatever happened after. Here was this extraordinary event, definitely outside the realm of ordinary human experience, that had just blown through my life, and I needed to understand it right down to my socks (as much as a body can understand such things), and validate it for myself: this did happen; this is what that was. Each day worse than the day before, accompanied by the diminishment of grace. I have heard it said lately that a well-seasoned rifle soldier is your basic functioning psychotic; as I say, we were not fun to be around. And all these years since, I cannot get around the fact that I was not simply a witness, but an integral, even dedicated, party to a very wrong thing.

Rifle soldiering, the downward path to wisdom to be sure.

By the way, it bears mention that our war in Vietnam was unwinnable. For the whole of my adult life there has circulated the bizarre contention that the United States could have won the thing—if only we had done this, or that, or the other; that wish list stretching to the horizon. Where did this idiot notion come from? Which Defense Department think-tank chuckle-head did *that* math? To say we could have won the war is the same as saying that we didn't fill our hearts with enough hate; didn't shoot enough Vietnamese down like dogs; didn't dispatch enough of their wounded with large enough caliber bullets to the head; didn't dump enough of their corpses in the bushy ditch-scrub like so many roadkills; didn't throw enough of them out of helicopters; didn't butt-fuck enough of their women; didn't Zippo enough of their hooches; didn't napalm or strafe or frag them hard enough; didn't poison enough acres of their woods and farmland with Agent Orange; didn't bomb them with enough B-52 air strikes ("whispering death," they called it) into small enough pieces far enough back into the Stone Age—the long, broad swaths of bomb craters the size of working Iowa farms.

At the end of the summer I had no idea what I wanted to do, but it sure wasn't driving a bus. I enrolled at Columbia College, a small arts school here in Chicago. The school had an open admissions policy—no SATs, no long drawn out selection process, no nothing; basically all you needed was a high school diploma and a checkbook. Since I had the GI Bill, I was money in the bank, and they were definitely glad to see me. I was one of the very few veterans among the students. I took a writing course because everybody knew it was a snap "A" and you didn't have to work; what the hell. But did I ever get surprised. The first night of class we introduced ourselves and said a little something about what we wanted to write about. This was just weeks after the convention, after a long hard summer of turmoil that stretched back into the early spring to Martin Luther King's murder and that free-for-all of grief when Westside went up in flames, and the city still twitched in that sweeping aftermath—not to mention the goddamn war—and everyone felt pretty edgy. More than one kid had been in the streets during the convention and wanted to write about that: running Lincoln Park through the rolling fog of nauseating tear gas and one step ahead of some cop thrashing at them with a billy club; getting hauled out of Henrotin Hospital emergency room by the hair, tossed in the back of a paddy wagon, and beat senseless; the drug scene on the streets of Old Town, and other such stories. I said that I had just come back from overseas and wanted to write about *that*, and the whole room just about sucked its breath and looked at me—you're one of *them?* And I looked back, thinking, yeah, I'm one of them, and if anybody wants to talk about it we can step out onto the fire escape. The teacher, however, was very interested; it turned out that he had been a medic during the Korean War. After we talked for a while, he pulled out a couple books and we read. The first was an excerpt from Jerzy Kosinski's *The Painted Bird*—the chapter about a miller at the dinner table taking a stew spoon and gouging out the eyes of a kid he catches eyeballing his plump little wife; the boy thrown out of the cottage, staggering around in the dark and screaming his head off with blood streaming down his cheeks. Then the teacher cracked open a thickish-looking book and started reading a story about a bunch of guys on a ship out in the middle of the ocean somewhere; somebody sights a whale and everybody jumps in these little boats they have, and rows after it; one of the boat crews finally run it down, harpoon it, and the mate (smoking a pipe all the while) kills it by stabbing it in the heart with a lance longer than the spread of your arms while the animal thrashes around spitting up great shots of gore through its spout hole; and finally when one guy says that the whale is dead, the mate takes the pipe out of his mouth, sprinkles the dead ashes in the bright bloody water, and says, yes, both pipes are smoked out.

You could have heard a pin drop in that room, and I'm thinking, *that's* a body-count story. I was never much of a scholar, much less a student of anything, and I asked, what story was that? Well, of course, it was Melville's *Moby-Dick*, and the teacher looked at me like I was some wandering alien just in off the street. Right then and there that novel went straight to the top of my list. The next week the teacher came up to me before class and said that if I wanted to write about the war that I should read these, and handed me copies of *The Iliad* and *War and Peace*. Now, I was one of those guys who had learned to read by moving his lips and following along with his finger (reading out loud always seemed to help), so it took me a solid year to read those three books. *The Iliad* is the absolute paradigm of the war story form, among other things. *War and Peace* is the whole story of a generation at war against one of the great military minds in all of European history; what a great yarn, and what a hell of a storyteller Tolstoy was. *Moby-Dick* has to be the greatest of all American novels, hands down, against which all other stories in our literature are compared, and for my money a number of stories have come close, but no cigars. Melville's novel is about a lot of things (including a deep understanding of and pleasure in American English and the tall tale as a story form, the same as Whitman and Twain), but it's also a shitty-job story; reading it, you get a keen appreciation why the passing of that work of slaughterhouse butchery is not mourned.

So, still sore and raw from the war, I began writing while the sour ambivalence of Americans was at its most touchy extreme—even among those people who welcomed me home with unconditional warmth and gratitude at my having arrived in one piece and without a scratch.

And I took my cue from Melville. Look at it this way: at one level being a soldier is just like any other work with its rules and results, punch-in and punch-out, make-work nonsense and shortcuts, and such. Do a job right and there is almost a physical satisfaction; but how can a soldier feel good about the work he does—combat as work simply cannot be satisfying in that way of ordinary wages-work. War produces an astonishing, pervasive ugliness, and that's all. To my mind a soldier's job is never done well, simply done with.

I would write about how the war worked; barracks language and barracks life; what the tracks were, what it was like to run the roads, and what it was like to plow your way through the woods, making your own road by knocking down trees one after another; how ambushes were supposed to go and what happened when an ambush went terribly wrong; the firefights and battles and what happened after; how the beer cans danced on the tables at the EM Club while ten miles away the B-52s dropped streams of bombs by the ton; the fuck you body-count stories of driving straight toward somebody with murder in your eyes (Homer's berserk), looking to run him down or shoot him dead with a ten or fifteen or twenty-round burst of the fifty; taking your life savings on your one-time-only R&R and fucking your brains out for a week in a Tokyo whorehouse; the excruciating nausea of diarrhea and that sloppy sound your body made as the shit poured out of you; the smooth-as-silk feeling that comes over you as you smoke a big fat tailor-made joint while walking in the dark of a morning to the mess hall for breakfast, and everything was just fucking fine for the rest of the day; the godawful grease trap shit smell of one-day, two-day, three-day-old corpses found leaning up against trees in the woods somewhere; the sore, numb feeling in your hands from working the fifty all night; the heavy, queer feeling that overtook your body as heat exhaustion overwhelmed you to the very ears; the foul spectacle of rows of lumpy, shining body bags shoved together like logs at a timber mill (the warm corpses of your friends still slimy with filthy sweat) in a monsoon downpour; the sour pathos of the exhausted, had-it-up-to-here medic while he tended to the umpteenth wounded guy screaming bloody murder as his very life poured out of him (nobody ever died quiet); that exquisite moment of dread that motorcyclists (bikers) called "seeing the movie" between the unavoidable crash and the instant later when you hit the pavement, sliding, and your whole fucking life passed before your very eyes; and finally that interminable, endless, nineteen-hour-long plane ride out of Tan Son Nhut to Wake to Honolulu to Travis when you sat, cramped and itchy, stoned about half out of your mind with grim joy, that intense coffee buzz in your head, nibbling on the pocketful of downers that the medic gave you for just this occasion, and you stare out that little window at the broad dark sea the whole trip.

Meanwhile I was reading everything that I could get my hands on—everything that came into my hands (many long afternoons in bookstores)—helter skelter, and without much regard for chronology; from Homer to Faulkner, Tolstoy to Burroughs, Chaucer to Dostoyevsky, the King James Bible to

Joyce, Fielding to Flaubert to Hemingway, Gogol to von Kleist to the Grimm Brothers, and a stack of war novels (Cervantes to James Jones) as high as you can reach. And, too, I ought to say that I was also reading bonehead English grammar books and dictionaries, because (as I said) I had never been much of a student; school had always seemed to be the place you had to stay until you were old enough to leave.

Yeah, I kept saying to myself, let's find out how to do this, and the intensity of the work matched the fierce intensity of my body and ambition; the stories just poured out regardless of who was listening. In fact, Edie has often said that during those first years I was wrapped as tight as spring wire (and she'd make a fist). So, I came to writing, the storyteller's craft, because I had a story to tell—a story that simply would not be denied and wasn't going away anytime soon—and not because I always had a burning desire to be a writer, and had started by keeping a journal since I was in Little League. And I would write about that whole long grind of a year's tour in a war zone—what that does to your body and your spirit, down here (touch just below your belly) where the Chinese say the ch'i resides.

The writers who emerged from Vietnam had the fresh example of writers from our fathers' generation of World War Two as well as the gritty, close-to-the-bone, walk-the-plank Beat writers of bust-out postwar American literature—those writers most accessible to us. We also had the considerable advantage of the broad permissions of the time for language and subject matter, and the simple honesty to *take* advantage; to leave nothing out; to thoroughly embrace Joseph Conrad's celebrated imperative to, above all, *make* you see (whether you want to or not); to *give* you a literal, visceral appreciation of the story (whether you wanted it or not). The result is that the blunt realism and frank barracks language of much of the literature to emerge from the Vietnam War leaves very little of the realities of a grunt's work to the imagination (as well as the war's most grotesque absurdities and moral obscenities). In *The Great War and Modern Memory*, literary scholar Paul Fussell drops the remark that the British soldier-writers of World War One struggled to get beyond their culture-bound sense of acceptable language, propriety and taste, and the overweening resistance of public sentimentality, to find a language and a *way* to tell their story—that hell-bound nightmare world of ditch squalor, wholesale bloodbath murder, and lunatic madness. The weeklong drumfire artillery barrages distinctly audible in the south of England when the wind was right. The night screams of thousands of wounded, men and animals both, after yet another pointless assault (the troops referring to the Battle of the Somme, for instance, as "The Great Fuck-up"); that war famous for its arrogant, "waste-of-good-infantry," flag-rank stupidities. And as that war moved from one year to the next and then the next, grinding on and on and *on* (so that the soldiers could well imagine the permanent stalemate of slaughter stretching into their middle age), the accumulating pall of stink from tens upon tens upon tens of thousands of corpses lingered over the topography for miles; it must have been hideous (in the spring the battlefields of the Great War are rich, still, with the aroma of iron and must). Is it any wonder that great numbers of the soldiers were driven insane? As Mr. Fussell says, the story of the Great War was less a problem of "linguistics" than of rhetoric—but that's the story of any war, mind you.

So it was with the writers who returned from Southeast Asia

with any part of their imaginations intact and the least humane sense of themselves not so much diminished as condensed; but where *do* you go to find the language and the way to tell the story? "I've been scaled," a GI once told Michael Herr en passant, "I'm smooth now."

I began writing my first novel, *Close Quarters*, in 1968, the straight-up fictionalized memoir of Philip Dosier. The one-year Vietnam combat tour was dramatically pat; it begins the first day and ends the last, and though Dosier and I share a good many things, I know a good deal about him, but he doesn't know a thing about me. The story was published in 1977.

That done, everyone I knew said that now that I had gotten this out of my system, I could go on to other writing—and stories are everywhere, mind you—but as far as I was concerned, the subject was far from finished. I turned right around and started work on my second Vietnam novel, *Paco's Story*, which is basically a ghost story about the sole survivor of a company massacre and what happened to him when he came back to the States. I began with the impulse to write a parody of *Naked Lunch*—wouldn't it be something to sustain that kind of energy for a couple hundred pages—and *Paco's Story* was published late in 1986. It turns out that within the tight genre of Vietnam War fiction (both American and Vietnamese), there is a broad vein of ghost stories, which tells you a good deal about that war as a human event.

I wrote those two books in an attempt to make clear that *this* is what awaits you—or something like—that the work of the war will transform you into something *you* don't recognize; that the inevitable reverberations of the war are irresistible and virtually irremediable; that *this* is what you make when you make war. (The reverberations still provoke body tics and shudders; long nights, still; extraordinary nightmares, vivid and precise, still; and otherwise, yet and still, a *severe* unease.) More than once when I first began writing about the war I got that look—how dare you tell those stories, how dare you use that language, how dare you represent that point of view—but I pretty much didn't give a sweet fuck what anyone thought one way or the other.

It is not the least of a body's impulse to write a thing down, to make it external to oneself, so that it may literally be sent out of the house. Some folks want to believe that the writing must have been a wonderful cathartic; "playing all those horror tapes," as one guy put it. But I have not found it so. Writing as therapy, God help us; the world doesn't work that way. Let's just say that both of my war novels were written out of deep bitterness; put another way, the impulse to tell the story of the war rose out of an undeniable authenticity of exhausted, smothered rage perhaps more bitter than tongue can tell. Someone once asked me why I wrote war novels, and I told him that writing novels was more elegant than a simple "Fuck you."

My war-year was like a nail in my head, like a corpse in my house, and I wanted it out, but for the longest time now, I have had the unshakable, melancholy understanding that the war will always be vividly present in me, a literal physical, palpable sensation. Aside from being a literary scholar, Paul Fussell is also an ex-lieutenant who fought in France and got a Purple Heart for his trouble. He once said that despite everything he is reminded of the war every month when his disability check arrives, and that he would always "look at the world through the eyes of a pissed-off infantryman." And I have to say, Mr.

Fussell, I could not have said it better or more plainly; does *any*one misunderstand his meaning?

In other words, there is something—a rasped timbre of the voice, a raw shiver down the back, a grind of the teeth—some *thing* at the tail end of a long, long night that, undeniably and try as I might, simply does not care. *Pissed-off*—the perfect word for "it."

AS I said, my younger brother Richard and I were drafted together, and went through Basic Training together. Afterward, Richard was sent to Fort Sill and the artillery, and then to Germany where he was assigned to a Pershing missile battery. We exchanged many letters. Fort Knox was a pain in the ass, I told him, but at least we had Louisville on weekends. Richard had to say that Germany just flat-out sucked. The work was dumb. The Pershings—what junk. If it ever came down to the button, Richard said, everybody knew that one-third of the missiles would simply fail; one-third would launch, then take off in the wrong direction; and one-third would actually fly right. Going to the field on maneuvers was *so* phony.

That fall, at Fort Knox, my name comes down on the levy for Vietnam; I get orders and a month's home leave; February.

The morning I am to leave for San Francisco, Richard calls to say that he's at the airport, on his way home. I am packed and ready to go, tricked out in my dress khakis when he arrives at the door with his crammed-full duffel bag and his Discharge in his hand. To say that we're surprised is putting it mildly. Welcome home, Richard, we are definitely glad to see you. He and I sit in our very small kitchen, not much bigger than a pantry;

me with my back to the sink and Richard leaning against the re-
frigerator; the window just there, and the yard still frozen solid.
And for two hours we have one of the most blunt conversations
I have ever had with another human being.

How the hell did you manage a Discharge?

Well, Richard gets to Germany, reports to his missile battery, and, to cut a long story short, he cannot bring himself to
go along with all the NATO garrison Army stuff and very forthrightly starts telling people off. He tells off the senior NCOs
and company officers; he tells off a couple chaplains and a couple psychiatrists: this whole enterprise is stupid; you are a fool,
and why should I go along with anything a fool has to say?
Richard wouldn't let it go, and wouldn't do anything they told
him. They finally put him to work in the motor pool to get
him out of their hair, painting trailers—one of those pointless
make-work jobs of which, in the Army, there is a vast abundance. His battery comes up for a live-fire drill, is brought
back to the U.S., and sent to White Sands Missile Range in New
Mexico to launch a Pershing just to see if they can make it go.
When they pass through Fort Dix, Richard is taken aside,
processed out of the Army, given his Discharge, and turned
loose.

Richard has done the thing that everyone I know wishes he
could do. I sit there and shake my head, admiring him; laughing. It was beautiful.

There are many stories of guys finagling an early out, as it
was called (pounding down a bag of sugar per day to blow out
your blood pressure, not wiping your ass for a month, speaking
in tongues, simply not showing up "for work," etc.); a truly rich
mythology. And of all the stories I've heard, my favorite is the

˹ible bicycle. The story goes that
˹g an invisible bicycle every-
˹barracks; walked it along when he
˹y; rode it to the mess hall and the PX
˹ the field for training. Nobody could talk
˹his commanding officer, not the shrinks, not
˹(who seem to be in charge of these things—cut it
˹be a man, suck it up and do your duty, blah-blah-
˹The guy simply persisted, and finally the Army threw up
˹hands and let him go. He walked out of the Admin. Building
with his Discharge under his arm, fetched his bicycle, took it
through the main gate, leaned it up against the fence, staying
with the gag right to the end, and walked away.

A beautiful story.

Well, Richard felt terrible that I was going overseas, and
when I finally, finally had to get up and leave for the airport or
be AWOL, all he could say was, "Don't get hurt." But Richard
was so distressed that during my whole war-year he could not
bring himself to write; letters, by the way, that I sorely missed,
but I didn't come to understand why he didn't write until a couple years later.

Soon after I arrived overseas, our youngest brother Philip
dropped out of high school and, since he was only seventeen,
our mother signed him into the Marine Corps. He volunteered
right away for Vietnam. And you know how *that* went—they
needed bodies in "the Nam," so the first day of Boot Camp the
drill sergeants harangued everyone like they've never been harangued in their lives until everyone volunteered for Vietnam
to "fight the Cong." Anyone who didn't step up right then and
there was taken out back and walloped until he saw the wisdom

of sucking it up and doing his duty—you *will* volun.
damn your eyes.

That spring I got a letter from our mother saying tha
had joined the Marines and was stationed somewhere alo.
DMZ (the demilitarized zone along the Ben Hai River, a.
17th Parallel). I put in for an emergency leave to go see him a
tell him to go home, but the Red Cross just took its sweet tim
approving it. They had to verify who I was, that we were, indeed, brothers, and that Philip was, indeed, in the Marine
Corps, up-country someplace. The process took three weeks,
and that delay is the source of my considerable enmity toward
the Red Cross all these years later; like the late historian and
ex-Marine William Manchester, I wouldn't be caught dead giving money to the Red Cross. By the time I get to Da Nang, Philip
has been wounded (a concussion upside the head from a near-
miss mortar round) and medevaced, has convalesced aboard
an offshore hospital ship, and has been sent home to Great
Lakes Naval Hospital in North Chicago, though he lived at
home. My mother wrote that he was "doing fine." And by the
time I returned home, Philip was mostly healed up and ready to
go back to active duty at Pearl Harbor. In 1969 he was ordered
back to Vietnam, and I told everyone, including Philip, that this
was a terrible idea. I contacted a lawyer at my grandfather's old
law firm (a retired Marine Corps lawyer). I told him bluntly
that I did not want my brother to go back to Vietnam, that I
wanted him to find out who I had to buy off and how much it
would cost, and to spare me his retired Marine Corps colonel
"duty, honor, country" lecture. By 1969 my family had ponied
up more than enough. There was, of course, nothing to be
done. Philip was shipped off, though my mother made certain

that Philip knew that I had tried to get him out of it. Philip was angry that I would interfere. What? I could not bring myself to write him during his entire, uneventful war-year; just as Richard had not been able to write me. I could not stand the thought of getting a letter back literally stamped *DECEASED* across his address. What an agony of waiting, and when Philip finally came home after a year (and without a scratch), he and I did not exchange a word for the better part of ten years.

He got married and had two daughters, then one day walked off to get a pack of smokes and never came back (the same as our oldest brother did many years before to his wife and daughter). No one I know has seen him since.

Over the years, I asked my mother many times what possessed her to sign him into the Marine Corps, and she couldn't bring herself to tell me; nor could I get her to tell me why she told Philip that I'd gone to see a lawyer. Right up until the year she died she would not speak of it, nor say anything about what it was like having him, recovering from his wounds, around the house.

And Philip? He never spoke of his time in Vietnam that I ever heard, but whatever happened to him during that second tour he was never able to overcome, was never able to get around, and it brought forth a violent rage (according to our mother and his young wife) that became a permanent fixture of his persona. During those years the family gatherings at Thanksgiving and Christmas were sharply, literally physically awkward; uncomfortable doesn't begin to cover it. When describing Philip, the brother I felt closest to, "rage" is not exactly apt, not large enough or subtle enough, but just now rage will have to do.

I am reminded of Philip every day because my son Preston greatly resembles him.

Meanwhile, Richard married and fathered four children, but he was always restless. Every couple of years they would sell the house and move. Then one year he divorced his wife (the kids well into their teens and early twenties), immediately re-married, and quickly had another daughter. He and his second wife abruptly separated in 1999 and he moved to Tulsa. One night he called me out of the blue and we talked about family, churchy stuff for an hour, but I heard not a hint of the trouble shortly to come. Several weeks later we got a call from the police department of a small town south of Tulsa; Edie took the call. Was she sitting down? The cop wouldn't tell her anything until she insisted that she was seated.

Well, there was no easy way to say it. My brother Richard had committed suicide. His body was found on a country dirt road just north of town.

Good God.

The funeral was an agony. Suicide is always an expression of remarkable hurt; always shameful and inexplicable, void of solace and utterly sad. The unmistakable rush of grief was numbing and raw, and would not go away. Richard's body, I should say, was the first corpse I had touched in better than thirty years; combing his hair with my fingers over the heavy cosmetic plug that covered the bullet hole—more intimate than we'd been since we were kids. This was worse than the death of my mother, who simply wasted down to nothing with dementia in 1992 (I had said good-bye to her years before); worse than the yearlong, lingering death by cancer of my father in 1977 (the same week that my first novel was published; how on earth do you grieve and celebrate at the same time?), the man squirming around on his bed as if he were lying on salt; worse by far than the news of a good friend of mine killed a week after I left Dau Tieng once and for all—Emerson Cole, a superb mechanic renowned for his grinning and fearless self-destructive nature, had swerved his truck to avoid a woman rushing across Highway #1 south of Trang Bang, flipped it off an embankment, and the deuce-and-a-half fell on his head.

My nephews were all in their twenties. The second son, the athlete, was devastated and furious, and insisted on driving down to see the town and the road—that place. I told him that I didn't want him going alone, so the day after the funeral we called the police down there to tell them we were coming, and took off. It turned out that my brother had not told any of his kids the family stories. So while my nephew drove his Camaro like a bat out of hell through Illinois, Missouri, and into Oklahoma, I talked a blue streak; telling him about our upbringing, the family origin stories, the characters and stars and misfits beginning with our one-armed, toothless grandfather (retired farmer and bar-fly raconteur), and ending with the story of Richard and me in the Army, and how his father got himself out before he got hurt.

We arrived Sunday morning, and it was clear from our conversation with the cops that nothing like this had happened in those parts for quite a while. Richard had bought a 9mm semi-automatic pistol at a hock shop in Tulsa, climbed into his Jeep Cherokee with a car cup filled with peppermint schnapps, driven south until he found a likely looking out-of-the-way place, parked and got out, and put a bullet in his head; the door of the Jeep was still open when his body was found. There was still a bit of schnapps in the cup set on the dashboard. Among the strange and grisly details of Richard's suicide was the blunt fact that when you put a 9mm to your head and pull the trigger, it is the muzzle flash rather than the slug that kills you. Then the cop led us to the place where Richard's body was found; my nephew insisted and I insisted with him. He needed to see this. That's right, son, your father did not just disappear off the face of the earth. The cop led us out of town, and then down the most ordinary of country dirt roads. We pulled over—nine-tenths of a mile from the blacktop, as he put it—and got out of our cars. He told us that the spot where he found Richard's body in the road next to his Jeep was around there somewhere, repeated his expressions of sympathy, wished us well, and drove away (it seemed like he couldn't get out of there fast enough).

Now, as we had driven up we passed a large, dark spot that looked like a six-quart brake fluid spill; that had to be the place (about a hundred paces back); what else could it be? So, we start walking and not twenty paces from the place we walk into a smell I remembered from thirty years before; that, and suddenly there were bugs in our faces, in our hair. We slow-walk around that large stain of blood in the rounded wheel tracks. I rattle off more stories; family vacations, the fishing trips, the time that Richard and I (and a couple other Basic Trainees) went AWOL from Fort Polk, grabbed a cab out to some whorehouse way off in the woods somewhere for a Sunday afternoon, and how this shrimpy ex–Arlington Park jockey got drunk and was the hit of the place—even the women were buying him drinks. The ditch at the side of the road is deep and thick with head-high bushy scrub and a profusion of wild sunflowers as big as the palm of your hand. Weed-grown pasture to one side of the road and a woodline to the other side; what a place to die. Richard, Richard, Richard, Richard, Richard—me shaking my head. My nephew wants to know what it was about this stretch of road that appealed to his father; wants to know why a 9mm; why the head; why he divorced his mother; why he married *that* woman; what the hell brought him to *this* place? Why, why, why? It is, I tell him, anybody's guess; what can you say?

Our war in Vietnam is now thirty years past, and, as we speak, more than a generation removed from our war grinding along in Iraq, but soldiers' work is always the same. As a veteran and a writer I am a little chagrined, yet and still, that the whole topic of the Vietnam War can not only get a discussion going, but can also get the hair up on the back of my neck; it is clear that there is much, still, to talk about. Since "Vietnam," several other wars have come our way, including Iraq and Afghanistan (as of this writing), and I don't know about you, but I have watched and been appalled by the horror-struck nonchalance with which we seem to enjoy them. We are fascinated and re-pelled simultaneously by the endless loop of televised imagery and skimpy narration, oiled with the patina of exaggerated pa-triotism that begins with the dusty, desert-bred bogeyman, travels clean through the bloody wrath of the Old Testament, and ends with those prickly little tingles in the scalp, the mois-tened eyes, and the grand old flag; everyone declared a "hero" just for showing up; love of country as religious experience. On television, at least, the war has been justified and prettified in a way that is truly pornographic.

Bad news from far away; what would we do without it? The

riveting and absorbed, self-righteous repulsion that the Bible refers to as "lust of the eye"; what compares? Who was it first said that nothing is more beautiful as someone else's firefight.

I expect we can say that whatever writer's reputation I enjoy is based on my two novels about our country's war in Vietnam, now three decades past. And after publication of the second, I thought (rightly enough) that I had written enough about that large and ghastly event, a benchmark of American history if there ever was one. But there remained, still, the itchy, undeniable sense of unfinished business between us Americans and the Vietnamese.

At bottom, behind and beneath everything else, was always the question, "Who *are* these guys?"

Then in 1990, my long-felt impulse to return to Vietnam came together with the opportunity. I was invited by the William Joiner Center at the University of Massachusetts in Boston to join a group of American writers going to Vietnam for a literary conference; the first of its kind, though there have been many since (Americans to Vietnam and vice versa). Among the writers and poets were Kevin Bowen, Philip Caputo, W. D. Erhart, Yusef Komunyakaa, Larry Rottmann, and Bruce Weigl.

The William Joiner Center for the Study of War and Its Social Consequences does just that, studies the phenomenon of war and its reverberating social aftermath. It is funded by the Commonwealth of Massachusetts, private foundations, and individual donations. William Joiner had been a Vietnam War—era Air Force enlisted man stationed in Connecticut loading shipments of Agent Orange, and later the university's director of Veterans' Affairs on the Boston campus until 1981, when he died of liver cancer long associated with that herbicide. The center was first established in 1982 and named in his memory. I have to say that it is very pleasing to me that the center is named for an ex-GI of the most ordinary kind, and a true brother of the blood. Initially the focus of concerned study at the center was our war in Southeast Asia, then those in Central and South America, and has rightly enough shifted to the small nations of the former Yugoslavia and now the Middle East and Afghanistan. There will, no doubt, never be a lack of subjects.

For two days in Hanoi we American writers met with Vietnamese writers from all over the country, including men and women we had likely fought. And for all the conversation and talk, one story stays with me.

During lunch one afternoon, I sat with Nguyen Lien, slight of build with elegant hands, professor of American literature at Hanoi University. I asked Lien what had he done during the war. He said his "task" during the American War (as the Vietnamese call it) was first to go to Beijing and learn English, then go to Moscow University and study American literature, *then* return to Hanoi, go out on the Ho Chi Minh Trail, and give classes and lectures in Whitman, Twain, London, Fitzgerald, and Hemingway, etc., to the troops moving south.

Then Professor Lien asked in all earnestness what Vietnamese literature had the United States Army taught *me* during the war. I pushed myself away from the table, threw my head back, and laughed good and loud, marveling. The honest guilelessness of the question could *only* provoke robust, derisive laughter from an American. The very idea of the United States Army giving anyone classes in literature of *any* sort is a mellow and ironic absurdity of the richest kind, and makes a body *need* to laugh out loud. I apologized for what could rightly be interpreted as mocking laughter and said that *I* would have been surprised if the United States Army had given *us* classes in Whitman, Twain, London, Fitzgerald, or Hemingway, much less any Vietnamese literature—which, by the way, goes back a thousand years and more. Professor Lien seemed surprised that Americans would not take the trouble to teach its soldiers something about the Vietnamese, but the fullest explanation would have taken the rest of the afternoon—and more.

Writing and poetry, "story" in the deepest cosmopolitan sense, have always been important to the Vietnamese. During the war, the People's Army newspaper, the *Quan Doi Nhan Dan,* routinely published poetry, stories, and literary essays along with the news. And the army's journal of arts and letters, the *Van Nghe Quan Doi,* published nothing *but.* The American armed forces newspaper, *Stars and Stripes* (published weekly, worldwide), never did that I ever heard; the thought would never occur to those people. North Vietnamese soldiers were encouraged to keep journals, write poems and songs, or whatever else came to mind, though more than one Vietnamese veteran has told me that paper and pens were impossible to come by. Americans, on the other hand, were forbidden by direct written orders from doing so—even from saving letters, for that matter—though no one gave *that* order any mind.

After the conference, we traveled to Haiphong, Da Nang, and Hue, then to Ho Chi Minh City, the Cu Chi Tunnels, and Tay Ninh near where I had soldiered around the Black Virgin Mountain.

And generally I discovered two things.

First of all, that the Vietnamese genuinely liked Americans, more so than they liked the Chinese (their traditional enemy), the French (who treated them like slaves in their own country), or the Soviets for that matter (the Vietnamese found them overbearing and arrogant; southern Vietnamese called them "Americans with no money"). Despite what we did to the Vietnamese during the war, their admiration and respect of us is heartfelt and genuine.

And second, by 1990 it was already a cliché among ex-GIs who traveled to Vietnam (Hanoi, no less) that we received a more warm welcome and expression of hospitality when we returned than we commonly received when we came home from the war. Part of this was, undoubtedly, the civilized politeness of a well-mannered host (never a thing to be taken lightly), but Hanoi poet Nguyen Quang Thieu, who graduated from high school the year the war ended and later studied English in Havana (of all places), put it more succinctly. "The Vietnamese people," he said, "were always able to make the distinction between the American people and the American government." And from what I have seen, this is true. This distinction between a people and a government is not one that Americans easily regard. As a nation we are too self-absorbed and self-important, simply too lazy (not to say arrogant), to take folks one at a time; much less to learn their language or read their literature. It is the root of our big-me selfish, schizophrenic pride.

Welcome, the Vietnamese of the north seemed to say; welcome back, said the Vietnamese of the south.

The road from Hanoi to Haiphong, Highway #5, parallels the single main line track of the French-built Vietnam Railway. At

Hanoi, the highway and the rail line share the Long Bien Bridge (the Doumer Bridge of French-colonial days) over the Red River as well as several other bridges along the way. Not only did I see passenger trains pulled by small diesel locomotives (the kind usually associated with light, local hauls and freight-yard work), but also steam locomotives hauling coal drags—what train guys refer to as "unit-train revenue service."

Steam locomotives in Vietnam?

During my war-year I had heard that Vietnam had a rail-road, but never saw it; the French-owned Michelin Rubber plantation at Dau Tieng shipped its processed latex down the Saigon River by barge straight to the docks in Saigon and didn't need the rails. Apparently at one time there was a rail line from Saigon straight north to Loc Ninh just east of us on Highway #13, but we never got that far east and I never saw it. That line moved through Loc Ninh and then to Phnom Penh in Cambodia. Most of the railroad in the south was destroyed during the war, in any case.

We flew to Da Nang, then drove north over the Hai Van Mountains to Hue, the imperial capital of Vietnam's last em-perors of the Nguyen dynasty. Highway #1 follows the rail line until it reaches the rising slopes of the mountains. The road eases left and rises toward the Hai Van Pass; the railroad crosses the highway and turns east, rising, to follow the coast; the mountains on one side and the broad stretch of the South China Sea on the other.

Larry Rottmann and I were both train buffs, and it was on that ride over the mountains, catching glimpses now and again of the slow-moving Da Nang–Hue market train inching its way along the coast, that we casually talked about coming back and riding the trains—wouldn't *that* be a kick.

Rottmann was friendly and outgoing, and helpful. For instance, at the Da Nang Airport we American writers and our Vietnamese hosts waited for our flight in a stuffy waiting room crammed with folks. It was a hot bright morning. Outside on the runway, a flight of powder blue Soviet Air Force MiGs took off one at a time with a long run to get up to speed, the engine roaring and each plane streaking past the terminal with that great crackling *whoosh* common to all overbuilt military aircraft, quickly getting airborne, and steeply climbing out over the South China Sea—out of sight in an instant. One plane, and another, and then another. They made quite a racket. Among all the folks in the terminal was a young couple with an infant, and the kid had clearly had it—even so early in the morning. The child cried and squirmed and fussed and bawled, and regardless of what the parents or the folks sitting around them tried to do, nothing worked. The kid's tantrum immediately got on everyone's nerves and wasn't going away anytime soon. Finally, Larry reached into his luggage, took out one of those kid-toy bottles of bubbles—you know, the liquid soap with the O-ring wand?—and made his way among the in-facing rows of scoop seats. He got down on his haunches in front of the parents and their baby, cracked open the bottle, drew out the soapy wand, and blew a string of bubbles into the air right past the kid's face. The kid had never seen such a thing and was instantly entranced and quieted; what splendid magic are bubbles to a child. The parents had never seen such a thing before. None of the Vietnamese in the room had ever seen such a thing before, and smiles of awe and gratitude filled the room. Half the people there wanted to buy Larry a drink.

He had been a first lieutenant in the 25th Division the same time I was there, and after he took a wound during the Tet Offensive was set to the task of writing a history of the 25th Division. His tour was also finished in March of 1968, and almost the minute he got back to the States he threw all his energy into organizing veterans for the primary campaign of Sen. Eugene McCarthy, the antiwar candidate challenging President Lyndon Johnson for the Democratic nomination. He joined the Vietnam Veterans Against the War, testified at the Winter Soldier investigation in Detroit (about war crimes and other abuses), and traveled to Washington to participate in Dewey Canyon III in the spring of 1971. He helped compile and edit the First Casualty Press anthologies *Winning Hearts and Minds* and *Free Fire Zone*, published in 1972, which were among the first books of poetry and prose by Vietnam veterans; both have become classics of the literature (and among the granddaddy's of us all). In the 1980s he began traveling back to Vietnam to sojourn the country from one end to the other, talking with anyone who would sit and talk, photographing everything he came across, and produced a number of documentaries, including "The Bicycle Doctors," about a husband, wife, and sons (all physicians) who worked at Hue Central Hospital, traveled to work and made house calls on their bicycles, and in the evening took out their doctors' knives and honed off the edges on whetstones.

What I realized during that first trip to Hanoi for the literary conference was that it was important for me to see the country at peace; to see ordinary folks leading ordinary lives without ordeal of the war. And if I began that first trip with a curiosity about who the Vietnamese were, the notion was also teased up about how did they do that? How did they defeat what history may declare the most powerful, if foolish, nation in the history of nations? Put another way: what did the Vietnamese know that we didn't?; did we know it once and forgot?; or did we never know it? In other words, what can the Vietnamese teach us?

I'll go back, hang around, and see what else there is to see. Besides, in the 1990s, Vietnam was an exotic place, especially for those of us who don't get out of the house much.

And since I'm a train buff, let's take the train; easy does it.

Here in the States if you "want to see the country," the best way is to get yourself a car with a good engine and solid brakes—at the very least—and take off down the interstate highways; that's what they're for.

In Vietnam the best way to get around is the railroad; stay on the train and it'll take you just about anywhere you want to go. Larry and I would start at Hanoi and work our way south to Ho Chi Minh City (called Saigon until 1975, when the war ended), lollygagging along, much like the Vietnamese did as their migration moved south from the Tonkin region of the Red River Valley, over the Hai Van Mountains (and the much-storied Hai Van Pass), and down to the Mekong River Delta—a movement, conquest, and migration that took about eight hundred years. Besides, I know just enough about how a railroad works to sit and talk with the train guys, and if there is one thing I learned from Chicago writer Studs Terkel it is that when you ask someone about their work, you get all these other stories.

So, Larry and I went back to ride the trains and I had a chance to take a look at a country I had never seen. I've been back several times in the last fifteen years, as time and money allowed, and the story that follows is an amalgam of those several return trips.

2

Good Old Bangkok

Why begin a trip to Vietnam in Bangkok? Well, the United States embargo; until 1994, going to Bangkok was by far the easiest roundabout way to pick up your visa.

What was the embargo?

As far as I've been able to understand, Lyndon Johnson's administration first imposed a trade embargo on North Vietnam in 1964 as a way of tightening the economic screws on Ho Chi Minh's government there; Americans were forbidden to trade in any way, shape, or form, and as a result nothing much moved into North Vietnam except from China or the Soviet Union, and not much of that until the later years of the war. In reality, the embargo made it all but impossible for the Vietnamese to get the most ordinary things from anywhere (everything from sewing needles to the most routine, first-aid-kit medical supplies).

By 1964 it became evident (from the American point of view) that the government of South Vietnam was not only unwilling, but also unable and all but incompetent to conduct its war against the North Vietnamese and the South Vietnamese guerrillas. In August the Johnson administration claimed that a couple of North Vietnamese gunboats attacked a couple of U.S. Navy destroyers; Congress was quickly stampeded into enacting the Gulf of Tonkin Resolution, greatly expanding President Johnson's war powers, and which the Congress would in later years come to sorely regret.* The bombing of the north began immediately, particularly the city of Vinh (about three hundred kilometers south of Hanoi and the jumping-off place for the Ho Chi Minh Trail).

Shortly thereafter the first American main force battle troops were dispatched to Vietnam; a couple of Marine Corps battalions sent to "guard" the airfield at Da Nang—a city with a superb harbor. This was followed shortly by more Marines, the U.S. Army's 1st Cavalry (Airmobile) Division, the 1st and 4th Infantry Divisions, and the rest (arrayed up and down the country south of the DMZ at the 17th Parallel). The 25th Division first arrived in late 1965.

The United States was finally, irrationally, and utterly

---

*At the time, it was claimed that North Vietnamese gunboats had made an unprovoked attack on the USS Maddox and the Turner Joy, two Navy destroyers. While the actual events remain ambiguous and contradictory, we may suggest to ourselves that this episode was among the first of a continuing "program" of our government's lies that would later include manipulated information of all kinds, such as the infamously exaggerated "body count" numbers, President Nixon's bombing of neutral Cambodia, and the like.

at war with the Vietnamese—contrary to abundant information, practical military considerations, and ordinary common sense.

Who was it first said that you cannot kill an idea with an army?

By the end of 1965 there were five American combat divisions in South Vietnam (including the 25th), 180,000 men all totaled, and the United States was spending upwards of $20 billion a year on the war.

Every year more troops were added, and by 1968, 554,000 Americans were stationed in South Vietnam. Before the Tet Offensive had ended, the big potato of Military Assistance Command, Vietnam (MACV), Gen. William Westmoreland, traveled to Washington (with his hat in his hand and a straight face, if long) telling one and all that we can whip them yet and asked for 200,000 more troops.

If there was a bigger fool in Southeast Asia than four-star General Westmoreland, a man with more punches on his ticket than the spring has buds, I have yet to hear his name. William Westmoreland is exceedingly tall, and he *looked* like a general in the same way that Ronald Reagan *looked* like a president. After his stint as commanding officer in Vietnam, Gen. William Westmoreland became the chairman of the Joint Chiefs of Staff at the Pentagon—the plumpest punch of all. And later still, he retired to Charleston, South Carolina, and slapped a bumper sticker on his clunker of a Buick: I Am A Vietnam Veteran; though I always thought *his* bumper sticker should have read: I Am *The* Vietnam Veteran. His participation in the Johnson administration's program of extraordinary lies—the routine inflation of "body count" numbers, the off-the-books bombings,

Phoenix Program assassinations, etc.—will always be his special shame.*

The war in Vietnam would drag on until 1975, our country's longest and most costly war; upwards of $165 billion. I have heard it said that each body-count corpse "cost" in the neighborhood of a quarter of a million dollars.

In January 1973, after years of diplomatic nit-picking, haggling, and stalling, direct American military involvement finally ended when the Americans and North Vietnamese signed the Paris Peace Accords. The South Vietnamese, knowing full well they were getting the shaft, refused to sign.

In March and April 1973, American and Vietnamese prisoners of war were exchanged.

The arrival at Clark Field in the Philippines of the American POWs from Hoa Lo Prison, famous as the "Hanoi Hilton," was broadcast live;† the planes arrived one behind the other. The former prisoners, mostly officers, mostly fliers shot down during bombing missions over the north, looked thin, prison-weary, and pale—as would anyone just out of prison (ask anyone who's done time in Mexico or Turkey or the old Soviet gulags,

---

*A number of American historians have criticized Vietnamese general Vo Nguyen Giap for wasting the lives of his soldiers in this or that ill-conceived campaign or battle. I bring this up only because it can just as easily be said that Gen. William Westmoreland wasted all of his. To paraphrase Samuel Clemens, William Westmoreland's stupidity would, by the least stretch, go around the world four times and tie.

† Hoa Lo Prison was named for the street outside the front gate; Hoa Lo means "the oven." The prison had been built by the French as the Maison Centrale to house ordinary criminals and the endless parade of revolutionary Vietnamese malcontents (of which there had been plenty over the years). For instance, General Giap's wife was arrested by the French in 1941, taken to Hoa Lo, imprisoned, and tortured, and according to American wartime intelligence was one day hung by her thumbs and simply beaten to death. Is it any wonder why General Giap had it in for the French?

or Atlanta or Leavenworth, for that matter). As each man came down the bit of stairs and stepped onto the tarmac, he saluted everything in sight, more glad to be home than tongue can tell.

The other meaningful exchange of prisoners, which was not broadcast and only slightly noted elsewhere, occurred at Loc Ninh, north of Saigon on Highway #13 near the Cambodian border (see map). The North Vietnamese and Viet Cong prisoners held by the South Vietnamese squatted on their heels in formation, and when the moment arrived for them to be repatriated, they stood up as one man (so the story goes), very deliberately took off their prison clothes, and walked naked to the waiting trucks. Prison is prison, to be sure, but pride is pride; not even our clothes would they take with them.

In late March 1973, the last of our troops departed, except those left behind in the guise of civilian mechanics and engineers to keep the South Vietnamese airplanes, helicopters, and other machinery working; MACV, the American military headquarters, was dismantled and sent home. Good, bad, or indifferent, South Vietnam was on its own. We still sent them abundant aid in cash, but the "loss" of the war—the reunification of Vietnam—was only a matter of time. It would not be our fault, at least on paper; which, for the lifers, was the only thing that counted.

The principal negotiators of the Paris Peace Accords, Secretary of State Henry Kissinger and diminutive Hanoi representative Le Duc Tho, were recipients of the Nobel Peace Prize that year. Simple, humble public servant Dr. Henry Kissinger (Ph.D.; Harvard, '54) did not even have the grace to

show up in Stockholm and accept the award into his own hand
from the King of Sweden; perhaps his ego would not fit through
the door of the aircraft that was to deliver him. Rather, he sent
a polite message of the blah-blah-blah boilerplated sort with
instructions that our ambassador read it; Dr. Henry Kissinger's
sarcasms about "peace" and "justice" take on an especially
fatuous edge when he quotes from William Faulkner's eloquently famous Nobel acceptance speech that "Man will not
merely endure, he will prevail." (Quoting someone else's Nobel speech, now *that's* lazy. No doubt Faulkner was rolling in
his grave.) Le Duc Tho, who sat in chair number five of the
Hanoi Politburo and whose specialty was southern affairs and
reunification, politely declined to attend as well, and pointedly
but diplomatically reminded the Nobel Committee and everyone else within earshot that in 1973 there was no peace for the
Vietnamese.

Then two years later, the American trade embargo was extended to the whole of Vietnam when the American-backed
government of the south finally collapsed from the weight of its
own corrupt and sloppy selfishness, and the country was re-
united under the leadership of Hanoi.

The bitter end went something like this: in March and
April 1975, one last large push by the North Vietnamese and
Viet Cong started a rout, and everyone scrambled to get out of
the way. (I have heard many stories of NVA soldiers chasing
ARVNs across the large grounds of Da Nang airport and having
to jump over thrown-off packs and rifles and other military
equipment as the ARVNs fled just as fast as their legs could

carry them.) By the end of April, Saigon was surrounded by five divisions of the People's Liberation Army of the National Liberation Front and legions of regular-issue North Vietnamese troops of the People's Army of Vietnam;* troops as well equipped and trained as we had been, armed with an ideology only too well suited to their gifts (as we and the southern Vietnamese, most assuredly, were not), and exquisitely motivated by several thousand years of fighting with foreigners. The Vietnamese had been polishing off foreigners for a good, long time and knew the drill.

The overall scheme was commanded by NVA general Van Tien Dung.

On April 21, 1975, South Vietnamese president Nguyen Van Thieu resigned and hightailed it for Taiwan with the wife and kids, along with all the cash, gold bars, and ordinary luggage they could haul. The presidency passed finally to Gen. Duong Van "Big" Minh, so called because he was almost six feet tall (uncommon for a Vietnamese) and more or less honest (also distinctly uncommon for a South Vietnamese general).

In 1963, General Minh headed the coup d'etat that had deposed and assassinated President Ngo Dinh Diem, a serious Catholic in a country of extremely serious Buddhists, a man renowned and deeply despised for his peevish mandarin zeal; the coup and murder just weeks before President John

---

* *The National Liberation Front (NLF), the revolution's guerrilla government in the south, was commonly referred to as Viet Cong; we called the guerrilla soldiers Victor Charlie, Mr. Charles, or simply Charlie. North Vietnam's military, the People's Army of Vietnam (PAVN), was called the North Vietnamese Army (NVA); they referred to themselves as* bo doi, *Vietnamese for soldier.*

Kennedy's own assassination.* Van Minh was then president for four months, had retired at the strong suggestion of another junta of ARVN generals, and had been cooling his heels ever since, virtually persona non grata, waiting for his country's call. He was South Vietnam's ace-in-the-hole of reconciliation with the revolution.†

On April 28, Van Minh was elected president almost by acclamation, by what was left of the South Vietnamese rubber-stamp legislature in the hope he could negotiate a cease-fire and truce, and then finagle some sort of coalition government with the NLF, whom Van Minh now called "our brothers from the other side." But the North Vietnamese and the southern National Liberation Front, which had been functioning as the shadow government, were in no mood to divvy up anything; Ulysses S. Grant would have understood completely. With the city surrounded—all but locked in—the jig was definitely up, and the only thing left for him to do was yield.

The final American withdrawal began on the twenty-ninth, and it was lock, stock, and barrel. The "secret signal" to commence the evacuation was a radio broadcast announcement that the temperature was "105 degrees and rising" followed immediately by Bing Crosby's 1942 smash-hit rendition of "White Christmas." Preposterous horsefeathers like this should give the ordinary American citizen a case of the giggles—Bing

---

* *Ngo Dinh Diem (pronounced "D'zee-em") is said to have coined the phrase "Viet Cong,"*
which in Vietnamese amounts to a racial slur.

† *ARVN, the acronym for Army of the Republic of Vietnam, was the overbroad generic term*
we used to describe anything connected with the South Vietnamese government. Say ARVN
fast ten times (it rhymes with Marvin), and you get an idea of how little regard we rank-
and-file soldiers had for most everything Vietnamese.

Crosby?; in 1975? "I'm dreaming of a white Christmas"; in April? Who thinks this stuff up? The Lord only knows what the Vietnamese thought, but then they'd been putting up with crackpot American brainstorms for what seemed time out of mind.

That evening, while one American helicopter after another arrived at the embassy (and a dozen other locations around the city) from offshore aircraft carriers for load after load of departing Americans and clout-heavy Vietnamese (first come, first served), the last of the embassy staff shoveled suprasensitive files and armfuls of one-hundred-dollar bills out of the safe and into a trash fire in the middle of the floor—the last of Ambassador Graham Martin's extravagant cash-money slush funds. It is meanly grotesque to imagine millions of dollars in American greenbacks going up in smoke. The embassy guys were still burning stacks of files and bundles of money at first light the next morning.

In war, nothing is more pathetic than the chaos of a skedaddle.

The final American leave-taking was so appallingly spineless, so clumsy and graceless, such a witless harum-scarum panic that the bodies of two Marines killed during the last hours of the evacuation at Tan Son Nhut Air Base were simply left behind; these along with military warehouses literally heaped with supplies, vast parking lots of equipment still waiting to be unpacked, cold American greenback cash by the bolt, the dog tag–making machines, and stacks and piles, reams and boxes and pallets of lists of AWOL GIs (for all anyone knows there are AWOL GIs in Vietnam *still*). The parade of helicopters to Saigon came and went all night long, swinging through an ad hoc ground-fire flak of liberated South Vietnamese rancor. The blunt, wholesale abandonment by well-connected Vietnamese and the Americans was not simple perfidy, but an agony, and an orgy of unambiguous betrayal. "They lied to us at the very end," said one American military officer, speaking of his own government (the guy apparently forgetting that it is intrinsic in the nature of governments to lie; perhaps it is what governments do best). Vietnam was, right to the end and still, a bungled tangle the American government had been picking at, tying and untying for twenty years and more.

By eight in the morning of April 3o, the last American helicopter was gone, once and for all (skied-up, we used to say).

How galling it must have been for the thousands of close-packed Vietnamese in the street, hysterical and desperate to flee (looking yet to the Americans for succor; pleading), when they saw that last handful of Americans climb the last paint ladder to that last bit of roof where the last Huey chopper waited to take them away.

What do you say to the man who is with you every day but the last day; indeed.

Saigon was completely cut off, doomed; Highway #1 east to Bien Hoa; Highway #13 north to Loc Ninh; Highway #1 to Cu Chi, Go Dau Ha, and Phnom Penh; Highway #4 to My Tho on the Mekong River (see map). For weeks, the 4 million Saigonese (3 million of whom had been rendered deliberately homeless by the war) had been told by the Americans and South Vietnamese government to expect a horrible and minute retribution for having the slightest association with the Americans; a bloodbath massacre the likes of which was beginning to unravel just that moment in Cambodia. Rumors swarmed of Da Nang policemen publicly beheaded; the Catholic bishop of Ban Me Thuot literally butchered; and three hundred people of that city beaten to death "with sticks." The headline in one of the last issues of *Stars and Stripes* to reach Saigon read: At Least a Million Vietnamese Will Be Slaughtered. No less a person than Secretary of State Henry Kissinger stood on his hind legs at a press conference and asserted that there would be "many executions."

With the Americans suddenly vanished as if snatched clean out of their shoes and the South Vietnamese armed forces all but melted into the very pavement like water spilled from a bucket, the Saigonese waited in terror for the presumed horrors to come. It was not difficult to conjure up knocks on doors in the middle of the night (sleepers shot in their beds), drumhead monkey trials and mass street executions, convoys of trucks heaped high with stiff and naked emaciated corpses showing only skinny shanks and filthy feet, and the brick by brick demolition of everything foreign (and therefore tainted).

Poised for days to regroup and await political developments, the Viet Cong and North Vietnamese moved on the city the morning of April 30. They expected the battle for Saigon to be long and hard (street-fighting is always the worst), but the lifers circulated a nifty slogan among the troops, "Once the bamboo is notched, one blow is enough to break it." You could imagine more than one Vietnamese veteran—the guys who had been at this a while—rolling their eyes and saying, "Yo! One notch! Ho! One last whack! I completely fucking forgot! Absolutely no fucking problem, sir!"* Nonetheless, the North Vietnamese and Viet Cong

---

* *I have heard the story from more than one Vietnamese veteran that as they marched in their battalions south along the Ho Chi Minh Trail, they'd meet groups of walking-wounded making their way north. The fucking new guys would sing and cheer, friendly-like, but the guys making their way north the best they knew how would reply, "Oh, fuck off."*

were surprised, even unprepared, at the ease with which they occupied the city, as if the Saigonese were simultaneously terrified
and impatient for an end to the war; paralyzed and relieved all at
once. It should be said that there were instances of what the
talking-head, armchair historians refer to as "bitter fighting" to
the last and many casualties—the Battle of Saigon was hardly a
walkover (especially around Tan Son Nhut; ask Bao Ninh)—but it
was also true that whole battalions of ARVN troops and entire
precincts of Saigon cops shed their uniforms and disappeared
into the city. I have heard the story that thousands of Basic Training troops at the recruiting center changed into civilian clothes
and went home, leaving neat ranks and files of uniforms, helmets, boots, and rifles behind—a weird and marvelous spectacle
that remained so for days. (It is, after all, not difficult to em-
pathize with meat-grinder draftees conscripted into an army
owned and operated by a power structure understood to be a
bunch of corrupt and grasping, hang-around-the-fort, get-rich-
quick losers.) To be sure, there were many conscientious and talented true believers whose patriotic altruism is to be honored
and respected, but let's not make a mistake here; the nation of
South Vietnam was an American invention of paranoid conven-
ience, and the southern oligarchy was definitely getting it while
the getting was good (as my grandfather used to say).

The immediate objectives for the Viet Cong and North Vietnamese troops were, of course, the major thoroughfares and
intersections, the bridges and riverfront docks, the various
headquarters and barracks, the police stations and radio/TV
broadcast facilities, the schools and universities, the banks and
ministries and other prominent buildings. Perhaps the most
symbolically dramatic and definitive event in the Battle of

Saigon, certainly the most patly photogenic, was the platoon of three tanks ordered to "take" Independence Palace; Doc Lap, a gray and hideous, five-story structure in downtown Saigon near the red brick Cathedral of the Virgin Mary and the French-built Central Post Office. It was here at Doc Lap that the South Vietnamese presidents had lived since before the Diem coup d'etat in 1963. The tankers' orders read: "Cross the Thi Nghe Bridge. Proceed straight ahead on Hong Thap Tu Street. Go seven blocks and turn left. Doc Lap is right in front of you."

Can't you just see it?

That clean and easy early first light of a Southeast Asian morning; barely enough to see. Strung out for miles along Highway #1 to Xuan Loc and beyond, the eastern phalanx of the People's Army rouses and cooks one last wartime meal of rice gruel, water spinach, and bitter tea. At 5:00 A.M. the whole army cranks it up and moves west toward Bien Hoa, the Saigon River, and the city (American helicopters still coming and going just as fast as they can load and leave). There are hundreds of tanks and trucks on the highway, thousands of troops on foot. Disorganized token resistance is "brushed aside." By eleven the tanks have reached the Thi Nghe Bridge over the river, and in no time the bridge is secured. Other platoons and companies and battalions of troops and tanks spread out toward their objectives. In its turn and with special care, not to say their hearts in their mouths, the designated platoon of three tanks crosses the river with every caution. For years these guys have been driving their Russian T-54s all over creation; they've busted jungle and camped, scouted and dug in, prowled and skulked, and have finally chased the ARVNs down Highway #1, elbowing past hordes of refugees, overtaking mobs of stragglers, and scooping up truckloads of abandoned American-made equipment at every mile.

The tanks crackle and squeak and grind across the bridge; can there be anything more ugly than the sound of tank treads on city pavement? It is the raucous and nasty, wracking gut-mean shriek of twentieth-century warfare. Easy does it, guys, let's not fuck this up; no bonehead moves, please. That familiar clammy-cold, dead-lug corkscrew of dread-fear tingles up the spine, goose flesh and frosty shivers; all it would take would be one lucky hit by one ARVN Boy Scout fighter jock with one five-hundred-pound bomb, and they could kiss it good-bye, bridge, tanks, and all. Today is *not* a good day to die. To be the last poor dumb son of a bitch to die on the last fucking day of the war is too dismal to reflect upon. The tankers ease clear of the bridge and move straight down Hong Thap Tu Street. Suddenly two ARVN tanks roll into the intersection, downrange, swinging their turrets around. There is a brief, sharp, virtually point-blank exchange of heavy, large-bore gunfire. The NVA gunners are better shots and soon the ARVN tanks are torched, but the burning hulks block the street. The splash of orange fire and greasy, diesel-black smoke; the blunt shock of the *booms* when the heavy ammunition inside begins to cook, the iron-colored ash that settles on the street and nearby rooftops as softly as leaves; the boil and drip of aluminum alloy armor plate; the crews inside dismembered, beheaded, eviscerated, rendered into offal, then burnt to a crisp—all this takes the briefest moment, you understand—bang, boom, *sloosh*; fry. The platoon turns left and immediately loses its way. Here we are, *so* close—almost literally a stone's throw. The platoon drives on, guessing. They finally stop and ask directions from the one thoroughly flabbergasted bystander in sight, a girl (some kid on a Honda scooter), gawking, more flummoxed than scared, eyeballing the three tanks and confounded that she has not been shot dead where she stands; Jesus, Mary, Joseph, may the Saints preserve us, but those things are huge! *("Troi oi!")** An interesting moment of negotiation unfolds: Young lady, which way to Independence Palace? The what? Independence Palace, which way? Independence Palace! Sure, everybody knows where *that* is. Next corner turn right, go straight one block. You cannot miss it.

There it is: the high wrought iron fence, acres of the best-looking piece of lawn in all of Southeast Asia, and the palace itself looking like something yanked right out of Disney World. Half a block in front of them is the grand front gate; chained and locked—what hope—as if that alone would prevent the downfall of the government. The lead tank, number 843 boldly stenciled on the turret, stops in the street and fires a shot clean over the place; we're here! It is the last high-explosive artillery round the platoon fires in the war. History does not record what goes through the minds of the people inside the building when they hear the shell sail close overhead (as serious a sound as any of them are likely to hear in *this* life). We may, however, presume that they expect the next round will come through the front door and straight down the hall. Then, so the story goes, the driver of the lead tank eases up to the gate. Of all the things this man has seen and done all these years, *this* is going to be a piece of cake. Just as he rolls up to the gate he guns it, the billowing surge of diesel exhaust blows straight into the air like

---

* *Pronounced "Choi oi," Vietnamese for "Wow," or the more earthy "Son of a bitch!"*

the whoosh of a bonfire when it catches all at once, and those wrought iron affairs snap clean off their hinges. Dust rises with a *fump!* as the tank slams those things to the asphalt. Several foreign photographers and journalists, taking pictures and jotting notes (and standing in everyone's way as if they were the most crucial aspect of the occasion), shout and point to the majestic entrance across the way. The drivers stomp on the gas and make a race for it. Yee-haw!

Fuck the lawn and the rhododendrons. The stone curb collapses like brittle candy and the tanks spread out and make that last easy dash straight across those last couple acres of grass, heading directly for the wide ceremonial steps. This is the place where many a South Vietnamese president (hero of one coup; pigeon to the next) has smiled big while glad-handing yet another big-potato American (arriving with softball questions, patronizing "talking points," and bags bursting with fresh, crisp American cash—a little something for your retirement, Mr. President). The photographers and journalists trot alongside the tanks, snapping shots—their spare cameras swinging from their necks. The tanks skid to a halt within an ace of the porch steps and overhanging eaves. The drivers slam the gearshifts into Park and stand up in their hatches, laughing, laughing, laughing—This is fucking sweet, brother; Welcome to fucking Saigon; Where's that fucking rice wine! One of the tankers, the platoon commander, the lieutenant, pops out of his hatch—excited, giddy (he wants to dance)—slaps a tightly folded flag under his arm, and whips out his pistol. Then he and his gunner, also with pistol in hand (everything locked and loaded, we're not out of the woods yet, bub), dismount and walk up the stairs straight into the building; let's put an end to this bullshit.

How odd it seems to be on foot.

Inside, Gen. Duong Van "Big" Minh, only lately president, walks at the head of a skittish entourage of rear-rank government ministers and back-office flunkies toward the two North Vietnamese tankers, moved by sheer human curiosity and accompanied by bowel-shrinking dread. Minh's only purpose just this moment is to end the killing once and for all with the formal capitulation of the government. For twenty years the best our money could buy.

Here is a genuine moment of Vietnamese history, but can you imagine anything more awkward? Consider: on the battlefield one of the most ticklish things to accomplish is to stand up, hands in the air, and surrender without getting hacked to death or butt-stroked and bayoneted or gut-shot by a dozen different guys. And when do you suppose was the last time General Minh stood in front of someone coming at him with a loaded pistol? Minh is exhausted with anxious grief and wants desperately to get the whole thing over, then get out of there as fast as he can.

In fact, his subsequent postwar "reeducation" is of the mildest sort, two days' "house arrest" in the palace, practically a vacation; other southern Vietnamese, not blessed with pull or a general's flag-rank star, are not so lucky; some will spend years, a decade and more, indeed, in the camps. It will be a lavish payback for all those generations of hot-box tiger cages and starvation rations, vicious tortures and wild beatings, car batteries wired to genitals and assholes and nipples—shooting bolts of lightning terror to the very ends of the toes, sparkling the eyeballs, smoking the hair, blowing the brain.

The tankers self-consciously kick off their sandals before they step onto the long and luscious yellow carpet and approach the new president and his escorts, but before "Big" Minh can say two words, the guy with the flag, still wearing his tanker's helmet, which resembles an old-fashioned leather football helmet with long earflaps and a chin strap, holding his pistol level and walking with clear purpose, tells Minh to *shit*can the small talk—Save it, pal—and asks the way to the roof.

The *what?* "Big" Minh believes he has misheard. Perhaps they mean the lavatory, but why would you want to go to the roof to take a whiz? (It is as if the preacher officiating at a society wedding were to suddenly launch into a dreamy monologue about the contrary philosophies of draw poker.) Then in sheer relief and baffled exasperation, Minh hooks his thumbs over his shoulder and has one of his petrified, bottom-rung, extremely junior assistant deputies show "our brothers from the other side" the way to the roof. The three of them take off down the wide palace hallway at a trot, leap up the stairs—taking them three at a time. The tankers begin to whoop like kids. The clerk, some pencil-necked drudge from Finance—a job his family bought him years ago—simply assumes he is going to be shot: I've been good, why me? Finally on the roof and out into the bright and hot, heavy noontime air above the city (and to the clerk's purgative and unadulterated relief), the tankers walk straight to the flagpole at the front of the building.

Magically, as if enchanted, the air above the palace grounds is thick with the whip and buzz of dragonflies.

The soldiers haul down the South Vietnamese flag (three red strips on a bright yellow field) with long, snappy reaches of their arms—wham! Then they raise the ensign of the Provisional Revolutionary Government (red and blue with a large yellow star, commonly called the VC flag), sharply salute with bold adolescent intensity, and a cheer rises from everyone standing within sight of it.

It is 12:15 P.M.

One could easily imagine such a cheer rising on Iwo Jima from the Marines scattered up and down the beach among the putrid, volcanic gravel dunes as half a dozen guys stood, finally, at the heights of Suribachi, tied the Stars and Stripes to a twenty-foot piece of pipe, and muscled it aloft; or the riproaring hurrah that rolled along the Carolina roadsides when one of General Sherman's supernumerary orderlies pounded hell-for-leather from one regimental headquarters to the next, shouting at the top of his lungs that Bobby Lee had finally surrendered at Appomattox, and one war-weary old lifer shouted back, "Well great goddamn, you're the son of a bitch we been *looking* for all these four years!"; or the positive *whoop-ha!* of extraordinary, deeply satisfying anger that rose from Sitting Bull's thousands as they rode down and killed every last one of General Custer's 7th Cavalry, which, years later, the Cheyenne chief Two Moon said took about as much time as a hungry man to eat his dinner—"Today *is* a good day to die!" the fuckingnew-guy youngsters shouted.

The clerk from Finance was never so tickled, nor grateful; jobs come and go, young man, but one's own sweet life, well now, that's something altogether different.

The foreign journalists, mostly from extremely neutral Europe, with a sprinkling of gutsy, stay-behind Americans, take plenty of pictures and scribble plenty of notes.

The war, a vicious agony that consumed the lives of two generations and more—an argument could be made that the revolution began the day the French showed up in the middle of the nineteenth century—the war, ladies and gentlemen, is ended; over; finished. Won or lost, we have lived to see the end of it; a victory all by itself (ask *any* rifle soldier). What next? Who knows, but they do know this: we are a country of small-hold peasant farmers and cottage-industry mom-and-pop shopkeepers; whatever happens after is going to be hard, and *hard* for the rest of our lives, but, once again, our country belongs to ourselves.

The Republic of South Vietnam, a literal creation of the United States Government and whose sovereignty dwelt only in the gut-cramped imaginations of the most powerful men in the world (a sovereign nation because no less a person than the President of the United States *said* it was sovereign), was instantly dissolved. In fact, soon after the tankers took to the palace roof with the victory flag, an official delegation of political cadre arrived, and when Duong Van "Big" Minh greeted them with the solemn presidential announcement that he was fully prepared to surrender the government, an NVA colonel was reported to have told Minh that he could not surrender what he no longer possessed.

Gen. Duong Van "Big" Minh had been president less than forty-eight hours.

And Vietnam, now called the Democratic Republic of Vietnam, was reunified top to bottom for the first time since the French arrived in all their splendidly ridiculous European pomposity over a hundred years before. It is altogether likely that no other modern European nation had a worse record of blunt colonial exploitation than did the French in Indochina, with the possible exception of the Belgians in the Congo. "Exploitation" is a loaded word, I know, but I've looked through the dictionary and cannot find a better one.

Soon after, Le Duc Tho entered the city for the first time in thirty years and sent a dispatch to his colleagues in Hanoi; a poem—

> Last night you didn't sleep, I know.
> The last battle began at dawn.
> You waited
> And followed minute by minute . . .
> [Now] the red flag stands forth
> To mark the entrance of our heroic and victorious army.
> . . . Ah, these tears shed for happiness,
> This joy savored only once in all one's life. . . .
> Uncle Ho's dream has become reality
> And he will sleep in peace.
> The sky today is splendid and infinitely serene.

Certainly not the most superlative poetry, but then when was the last time an American sent a battle report in the form of a poem? Perhaps the last recorded instance of superlative American flag-rank poetry, a one-word haiku if there is such a thing, was penned Christmas 1944 by Brig. Gen. Anthony McAuliffe of the 101st Airborne at Bastogne during the Battle of the Bulge. Hopelessly and solidly surrounded, the Germans demanded his surrender, and General McAuliffe responded with a note that said, simply, "Nuts," which everyone but the Germans understood to mean "Fuck you."*

Saigon was renamed Ho Chi Minh City. "Uncle" Ho had died in 1969, rightly regarded as one of a handful of great Vietnamese patriots.

The movement of twentieth-century history dictated an end of European colonialism, helped along by the economic and cultural exhaustion of two "world" wars before half the century had passed. Ho Chi Minh may have been the bluntest of communists, but he was also a stone-fucking nationalist, a superbly gifted and pragmatic politician, and a native of a country that first began fighting foreign invasions and occupations two hundred years before the birth of Christ; most notably against the Chinese, Vietnam's traditional enemy. He was, after all, the guy who, in 1945 when the Vietnamese had to swallow whole the return of French colonialism, told the assembled crowds with what could only have been considerable, exasperated chagrin that it was better to smell French shit for ten years than to eat Chinese shit for a thousand; we may be sure that none of the Vietnamese within the sound of his voice misunderstood. After all, Ho Chi Minh had expatriated himself for

---

* *The story has circulated ever since that General McAuliffe did indeed respond "Fuck you," but that the public relations colonels at General Eisenhower's headquarters could not, of course, allow that blunt phrase to appear in the home-front family-oriented newspapers; though it is sure that all the lifers then (and since) got a good, hearty laugh from it—surrounded, out of ammo, freezing to death, and generally dying like flies—Fuck you, indeed; har-har-har, those "Screaming Eagle" Airborne guys—that's balls. (Down through the years many a veteran would say to his wife on many a cold, cold night, "It's cold alright, dear, but thank God at least it ain't fucking Bastogne.")*

thirty years to learn the way of the world, as the saying goes, and returned to Vietnam with an ideology only too well suited to the Vietnamese spirit. When the French returned to Hanoi and the Viet Minh left for the mountains north of Hanoi, the revolutionary war of liberation began in earnest.

Even so, everyone still calls the city Saigon.

The program of North Vietnamese bloodbath reprisals that everyone expected and feared did not occur, though the "reeducation camps" were no small thing. In fact, in the days immediately after the war it was not uncommon to see North Vietnamese soldiers, generals included, scouring the neighborhoods with a decades-old address in hand, looking for family not seen since the country's partition in 1954.

The Vietnamese turned inward and began rebuilding, and, like many another people after such godawful devastation and the upheaval of war, looked to rediscover who they were; never a simple, straightforward matter.

And the American embargo?

The administration of Gerald Ford (who became president after Richard M. Nixon resigned in abject disgrace rather than be impeached for "high crimes and misdemeanors" of which he was certain to be convicted), with the final, piddling arrogance common to all tyrants, extended the trade embargo to the whole of Vietnam. United States companies could not trade with the whole of Vietnam in any way, shape, or form. Vietnam entered a decade of isolation as fiercely self-exclusive as it was, apparently, sadly necessary; trading only with neutral countries and other socialist governments.

American law was quickly enacted to make it all but impossible for Vietnam to get loans and grants-in-aid from the International Monetary Fund, the Asian Development Fund, or the World Bank (where Robert McNamara was making mischief after he left the Johnson cabinet as secretary of defense in 1968). And, as well, our government made sure to squelch Vietnam's membership in the United Nations until after 1980.

So it stood, and for a decade and more life was pretty grim from one end of Vietnam to the other; during the wars, life in the north had always been hand-to-mouth, but with the collapse of the artificial made-in-America economy of the south, the whole country suffered alike. During that time, Vietnam's biggest export was the scrap metal of leftover war junk; nobody gooses up the juice of a local economy like Americans, and nobody leaves more "stuff" behind than an American army. Meanwhile, many Americans and Vietnamese started working for a normalization of relations; forward-looking and progressive scholars, ex-GIs, and folks from organizations like the American Friends Service Committee (the Quakers).

Then in early 1987, *doi moi* (the "economic restructuring" that swept through many a socialist country) became policy in Vietnam, and the possibility of a normal relationship seemed close. In the United States, however, the matter was muddled by the issue of unaccounted-for Americans still listed as missing-in-action, the MIAs; the discussion made bizarre by hard-assed conservatives, irked to distraction over losing the war, who absolutely insisted that there were, *still*, Americans being held prisoner, and when *that* failed, demanded with shrill argument that there be a full accounting of the MIAs and a return of every last missing man's remains;* they even had a flag. Let it be said that if at the war's end there were 2,500 American MIAs on the books, the Vietnamese were missing 300,000—give or take.

Some Vietnamese, I can assure you, will never be found. I know, because I saw them transformed into spray.

Finally, in 1994, when it was absolutely safe to end the embargo, President Clinton called the whole thing off. The very day the embargo ended, Pepsi cola handed out free cans of "product" in Hanoi, but the Vietnamese never needed Pepsi cola half as much as they needed penicillin; still do. A State Department spokesman told me the end of the embargo meant that "You could do anything in Vietnam you can do in France." Which I'd say covers a lot of territory.

In 1995 the United States and Vietnam granted each other diplomatic recognition and agreed to exchange ambassadors. Our first ambassador to Hanoi, postwar or otherwise, was Douglas "Pete" Peterson. An Air Force pilot during the war, Mr. Peterson was shot down in 1966 near the village of An Doai in the Red River Delta east of Hanoi on his sixty-seventh sortie over the north and spent six and a half years as a prisoner of the war; at the time of his appointment he was sitting in the House of Representatives, where he was congressman for a district in Florida. Douglas Peterson was a canny, not to say clever, choice; who was going to argue with the credentials of the Honorable Mr. Douglas Peterson? It was altogether likely that he was the

* As of this writing there are 1,142 unresolved cases, according to the Joint Task Force—Full Accounting Project, which is the outfit active in the search and recovery of whatever remains of unaccounted-for Americans, working out of our embassy in Hanoi.

only man willing to take the job who (in 1995) the Clinton administration could parade in front of Jesse Helms, the loose-
cannon dufus chairman of the Senate Foreign Relations
Committee, and obtain approval.

The arrogant and small-minded Senator Helms was a genuine bumpkin with a stingy, backward imagination who hailed
from a region of the country that otherwise produced intelligent minds. He had a head like a great shaggy marble, the rumpled, disorganized look of a man who spits and drools when
he's drinking, and could palaver and pound sand with the best
chuckleheaded windbags of our long and glorious congres-
sional history. His skill at political sandbagging was second to
none; some experts flatly assert that he was the Charles de
Gaulle of sandbags. To think that this man was chairman of
*any*thing boggles the mind.

The Honorable Mr. Helms did not like the first thing about
President William Jefferson Clinton (who out-windbagged *and*
sandbagged everyone)—this years before President Clinton's
famous Oval Office blow jobs brought him before the congres-
sional bar of impeachment (one of those extremes of crackerjack silliness and rhetorical moral outrage our legislature is
famous for, God help us). So, Ambassador Peterson's confir-
mation dragged on for months; the man finally took up his post
in May 1997.

One year later, Ambassador Peterson married Vi Le, twenty
years his junior. Ms. Vi Le was an expatriate Vietnamese with
an Australian citizenship, born in Saigon after her parents re-
located there from Hanoi in 1954. She grew up in Hong Kong
and Australia, and was working in Hanoi as Australia's trade
representative when she met Mr. Peterson at a reception at the

Israeli embassy. They conducted a very public courtship. Many Vietnamese regarded the marriage as a telling symbol of reconciliation between Americans and Vietnamese.

Before the Vietnamese government declared *doi moi* in 1987, foreign visitors were not encouraged. Even with a visa, traveling there was a complicated and frustrating, all but impossible business, and Americans in Vietnam were singularly rare; perhaps a hardy handful. After *doi moi*, Americans began returning, first in ones and twos, then in small groups, then in more and larger groups. You sent your visa application (with photo attached) through the mail to your sponsor (in my case the Vietnam Writers Association in Hanoi), were notified of approval (also by mail), and traveled to Bangkok to pick up your visa. The flight from there to Hanoi was a matter of an hour and a half—about the distance from Chicago to New York. The sheer logistics of the process was exasperatingly slow, which was one reason why the 1990 conference of American and Vietnamese writers took years to arrange.

And even if there were no snafus, a two-day stopover in Bangkok was not unusual. Nowadays, what with the Vietnamese embassy in Washington, getting a ninety-day tourist visa is routine.

Very westernized, Bangkok is; gaudy and earthy to boot, right down to the horrible air pollution from bumper-to-bumper traffic, even at midnight on the freeway. During the daylight hours it's much worse—the boulevards awash with foot traffic, peddlers, brazier-equipped four-stool fast-food sidewalk cafés, scooters, *tok-toks* (three-wheeled open-air taxis), diesel buses hauling commuters from the countryside (a two-hour ride is not uncommon), Mercedeses and Hondas and Toyotas, and trucks of all descriptions as well as mint-condition Chevys and Jeeps, and other makes of downright mysterious manufacture.

Now, being stuck in Bangkok for a couple days on your way to Vietnam is not *that* much of an inconvenience. It is a popular tourist destination and one of the most exotic capital cities of the world; a beautiful place to kick back and let your jet lag mellow down. The Thais are world renowned for their hospitality; "The Land of the Big Smile" all the brochures will tell you. During the war, of course, Bangkok was *the* favorite R&R city among American GIs, famous as the Whorehouse Capital of the World; if you couldn't get laid in Bangkok, you weren't going to get laid anywhere. Every Bangkok R&R story I ever heard amounted to a culinary and sexual rampage; the young Thai women only too glad to oblige.

The American military has always been proud to say that it provided "our boys" on R&R with *every* "entertainment," beginning with a very thorough list of whorehouses awaiting you in your R&R city of choice. The women, not any older than we were, would be waiting at the bar; take your fucking pick, soldier.

From there, the in-country Vietnam, USO-sponsored "entertainments" worked backward to those dreadful all-girl garage bands tricked out with flippy miniskirts (the first we'd seen), white go-go boots, and Barbie Doll hairdos; the Filipino trios that backed the all-but-naked "singers"; and the jive-ass

Hong Kong rock 'n' roll bands that played and sang by rote. Then came the traveling troupes of *Playboy* bunnies with their skippy feet, their gee-willickers, red-blooded American, girlie-next-door porcelain-doll tits and ass who swooped around the countryside from one shit-hole firebase to the next in a 1st Air Cav Huey, flashing whiplash smirks, look-but-don't-touch peek-a-boo cleavage, and thunder thighs. Next on the list were the truly arrogant delegations of off-season professional athletes (future Hall-of-Famers packing certified 4-F knees, out-the-bubble psych profiles, wink-and-a-nod deferments, or phony, no-show National Guard "jobs"). They wandered here and there and yonder, spreading facetious big-league public relations cheer around the big-name division evac hospitals; these PR jocks photo-opped and chatted up and glad-handed the post-op litter-wounded who were in just as good physical shape as they—except for the trauma amputations, the through-and-through sucking chest wounds, the third-degree white phosphorus burns on their backs, and the cold-sweat malaria shakes. Then there was "Up with People," an ensemble of perhaps a hundred "boys" and "girls" that toured military establishments the whole world 'round; even more clueless than *Playboy* broads or some of the Red Cross women. The "Up with People" people performed godawful amateurish "showstoppers" from classic Broadway musicals and swinging medleys of beach-blanket-movie love songs. They even had a peppy, circusy rendition of *The Mickey Mouse Club* theme song—this years before the closing scene of Kubrick's *Full Metal Jacket*. We looked at these people with blinking astonishment as they swooped and whooped around the stage—

"What the fuck is *this?*"—while we sucked down our brewskis and smoked our OJs, laughing with our whole bodies until tears came to our eyes.

But the irremediable worst of the dipshit in-country Vietnam "entertainments" was the annual Christmas show organized around Bob Hope—that grandiose, stuffy old hack. The world-famous *Bob Hope Christmas Special* (filmed for one of the television networks and broadcast later) was a veritable Hollywood Canteen parody of well-cured ham. Bob Hope's shameless self-promotion and Booster-Club after-dinner blather of jerkwater Jody-jokes made the very eyes of our asses sore. What bullshit. The very lameness of his primetime, mock-titillating Chamber-of-Commerce monologues (pushing the war as "product"), the very wiggliness of the no-name, over-the-hill celebrity bimbos he trotted out, the very absurdity of dumpy, buffoonish band-leader Jerry Colonna with his Cookie-Monster google eyes and *bandito* mustache, and the be-boppity, pickup ensemble of musician-union hacks were better suited to World War Two home-front bond drives—or primetime network television (the war's going just fine, folks; see, we got Bob Hope!). It was worse than an amateur circus, but the lifers couldn't get enough of him.

This nonsense rendered us speechless with wonder, but was no goofier than anything else the Americans dumped on the Vietnamese. As a matter of anecdotal curiosity, it would be interesting to know what the Vietnamese had to say back of their hands about all this big-nosed American nonsense.

These days, of course, Bangkok is the sexually transmitted disease capital of the world—good old clap, syphilis, herpes, chlamydia, HIV and AIDS, and other such foul miseries. Only an absolute fool on a once-in-a-lifetime, pedal-to-the-metal, what-the-hell-Bubba, fire-eating binge (not unlike an infantryman's R&R) would horse around in Bangkok. You would have to have rocks in your head. Though prostitution is strictly illegal in Thailand, no one pays *that* law any mind, and all-male tours from Europe and Asia can be seen crawling all over the central, strictly tourist part of the city; paying top dollar to sleep with (presumably) healthy and exotically attractive young women. I read in an article in *Ms.* magazine a number of years ago that by the time Thais reach twenty-five years of age, one-fourth of the women and one-fifth of the men have worked as prostitutes. Ordinarily the Thais are the pleasantest, perhaps the most beautiful people in the world, who take their Buddhism and their king very seriously, but life must be pretty discouraging in the countryside away from the tourist spots. We may assume, however, that the Thais keep smiling.

It was the better part of 1:00 A.M. when Larry Rottmann and I arrived in Bangkok and got to the (I am not kidding) Manhattan Hotel, near Sukhumvit Road. The hotel is within walking distance of embassy row on Wireless Road as well as such dynamite nightspots as the Edelweiss Bar, the Butterfly, the Susie Wong, and the Darling-Is-Feeling Massage Parlor, among others, where the women wear large laminated numbers pinned to their halter tops, and the weary traveler can get his ashes hauled in leisure and with considerable flourish, and then, of course, turn around and get his ashes hauled again. Our first trip with the writers in 1990 coincided with the World Cup soccer matches in Italy, and in the lobby were stacks of "Special Editions" of the *International Herald* filled with up-to-the-minute soccer news. Alongside these were stacks of slick-paper advertising tabloids hawking "special" shows, escort services, disco bars, and massage parlors—both traditional and contemporary (they'd be right over; they had everything but 800 numbers).

The next morning Larry and I moseyed down to the Vietnamese embassy to pick up our visas. Ironically, it's two doors down from the extravagantly secured American embassy compound, where you get frisked up one side and down the other, and then are given a stern cop-lecture from some bratty little weasel (working the cushiest of cushy summer jobs) who turned out to be the ambassador's kid.

We sauntered up to the Vietnamese embassy and walked in the modest little office just off the street. Behind the glass partition was a very plain office where a small woman sat in front of a disorganized stack of very large, canvas-covered ledgers—the sort you might see in a film of a Dickens novel. Spread around the room behind her were several ordinary gray-metal office desks. When our turn came, we slid our passports under the glass and said our names out loud. The little Vietnamese woman looked at our pictures, then at us, and checked our names. She then moved aside two or three of those huge, thick ledgers, set the appropriate volume squarely in front of her, opened the cover, and turned each page with a large and deliberate, casual gesture of her whole arm. At last she came to the page of current entries; we could see today's date upside down through the glass. With our passports spread out like tarot cards, she laid a finger on the page, and *bingo*, our names were right at the top.

She looked up and smiled, and we all agreed this was a very good sign. Then she put down the passports, went over to another desk, and very deliberately unlocked a drawer. Here was a modest stack of perhaps a dozen visa applications (with attached photos). She bent over and flipped through the stack with her fingers, then straightened up with a puzzled look on her face. She came back to the window, fetched our passports, hauled out the stack of applications (with attached photos), and went through the whole collection again, comparing the photos.

Then she made a very long face.

Uh-oh.

Well, I am sorry, gentlemen, but your visas are not here. I do not understand it. No, I do not understand this at all. Your names are certainly in the book, and that is as it should be, but your approved visa applications (with attached photos) are not in "the drawer," said she—that was the gist of it, anyway.

Without missing a beat, Larry and I reached into our camera bags, whipped out duplicates of the visa applications we had sent in months ago (with photos attached), brought along "just in case" there was a slip-up just like this, and gave them to her. All she would have to do is write in some numbers and we'd be in business.

"No problem," said she. Come back this afternoon around three (after siesta). "No problem," the international signifier of small-time glitches—you hear it everywhere.

I tell you all this to illustrate that until recent years everything, and I do mean everything, in Vietnam was done long-hand and with a studied (that is, cogitated) air of thoroughness. It is the sort of easygoing, move-slow-and-keep-to-the-shade pace common to every hot-weather, Third World culture (where no one is paid what they're worth); just the sort of calm easy-does-it touch that takes some getting used to and drives busy-busy, push-push, git-git-git Americans up a wall. Larry and I went back to the hotel for a jet-lag nap, a long lunch, and a bit of shopping, and when we returned to the little Vietnamese woman, our visas were out and on her desk; everything now copacetic. She whipped through the simple bureaucratic ceremony of stamping our passports, writing the dates, scribbling the initials, and polishing the whole thing off with a snappy bang of the official seal. Ah.

The next morning we left for the airport. Once we got through the routine chaos of the airline check-in, past the sleepy airport cops and customs guys, and out into the waiting area of the International Terminal, we sat among the rows of fiberglass scoop chairs with our legs over our luggage and waited for our flight. For those of us who don't get out of the house much, the destination board at Bangkok has to be one of the most exotic in the world: Amsterdam and Athens, Barcelona and Beijing, Delhi and Dubai, Jeddah and Karachi and Katmandu, Moscow and Muscat, Penang and Phuket, Rangoon and Rome, Sydney and Singapore and Zurich (and many, many places in between).

And there were all kinds of people and all manner of clothes: Indian saris (probably the most beautiful dresses in the world) and sandals, wool caps and serge trousers and walked-to-death Hush Puppies, business suits and regimental

ties and spit-shined wing tips, Danish "hippies" with Birken-
stocks and lavender socks, Japanese businessmen hauling
double bagfuls of duty-free booze, Sikh beards and huge tur-
bans, Italian mustaches, Swedish ponytails, and disheveled
Frenchmen with hair that looked like haystacks in a fit.

Sitting there, waiting for our flight to Hanoi, I could not
help but think of the contrast of all those years before.

In March of 1967, I was ordered overseas and reported to
the replacement depot at Oakland Army Terminal (what GIs
have called a repple depple since World War Two), and waited
for my turn with hundreds of others in the transient barracks,
a huge windowless warehouse. Inside were many, many rows of
close-packed double racks; many, many troops dressed in
their cleanest, crispest khaki uniforms, buffed-out brass, and
spit-shined shoes—all of which faded fast (you slept in your
clothes). The building positively smelled of bodies and Brasso,
was never quiet, there was absolutely nothing to do, the lights
that hung down through the rafters were never turned off, and
we were called to the door by the planeload (in another life it
would have been boxcars).

I have to say with no small chagrin that I have absolutely no
recollection of anything that happened during that score of
hours we were in the air, bound for Tan Son Nhut Air Base. I
could not tell you the places we stopped; cannot recall that we
ate; do not remember if I ever went to the lavatory. It was as if I
were already dead—distracted; numb—to everything, including
memory. Not easy to be a young soldier; very much like prison,
I think.

But I do recall that two things kept going through my mind. First was all those rainy-day gym classes of dodgeball when I was in grammar school; a "body-count" game if there ever was one. The coach would haul out this big bag of cheap, thick-rubber kick balls; the classes split into two teams—one at each end of the gymnasium. He'd set the dodgeballs in the jump-ball circle at midcourt, then he'd give a mighty blow on his Acme Thunderer whistle, and forty or fifty kids would rush in, try to grab a ball, and commence throwing them back and forth at each other. The rules were simple: no going beyond half court; if you threw a ball and hit someone on the fly, he was out; if you threw a ball and he caught it on the fly, you were out. On the plane I kept mulling over the extremely discouraging fact that I had never lasted very long at these games—always among the first to get whacked (greased, we troopers called it). In other words, my ass was grass.

And second, back at Fort Knox in the late summer and fall of 1966, large numbers of "fuck you" tankers began transfer-ring back from armored cavalry outfits. There was a vacant and desultory, dusky look about them that didn't have anything to do with the color of their skin or their leather-assed Southeast Asian tan. To a man they were clearly fed up, but in deep and obscene, not-so-subtle ways. It was as if they were on the verge of rearing back and howling the curse that had been bubbling up in them all that year; as if they possessed an intensely grim, harshly whetted bitterness that they now turned inward; never a good place to point such a thing, regardless of its timbre or velocity. They did not simply have "an attitude," it was as if they had no attitude at all. For instance, one Saturday afternoon the colonel had all five companies of the battalion out on the regi-

mental drill field for a grand, dismounted review (I do not remember for what occasion); even the officers' wives showed
up. It was raining one of those steady, chilly, early autumn
Kentucky rains. The officers and their women stood high and
dry under the broad awnings of the reviewing stand. The battalion stood in formation by company for the better part of an
hour, waiting for the weather to clear, but the rain never did let
up, and the colonel finally called it off. Every man in the battalion was wet clean through. Not long after, in my barracks' over-
crowded squad bay of double bunks, the colonel came to the
door with the sergeant major. Someone called us to attention,
and an instant later everyone stood briskly attentive while the
colonel thanked us for standing in the rain; as close to an apol-
ogy, I suppose, as his rank permitted. The tankers, however,
sat on their bunks and footlockers (stewing in their sopping-
wet clothes) and didn't even turn around, much less stand up.
The colonel chose not to notice, turned on his heel, and left,
but the sergeant major lingered in the foyer between the boiler
room and the latrine and laid into those guys like nobody's
business; the tankers just sat, eyeballing the sergeant major
with brusque, crusty detachment, and let him have his little
say. They had had it up to the very ears with parade-ground
chickenshit from brain-dead lifers like him and sincerely did
not care; aggressively and intensely did not. Most of them
drank morning, noon, and night; we watched one man go from
staff sergeant (E-6) to nothing flat in six weeks. Some talked
about the war and some did not. They poured out their pay on
big, fast cars and large, loud and ugly motorcycles; and the
minute they had their Discharges in their hands (the precious
DD-214) they whipped through the gate, swept out onto US

31W (called by us "Thirty-one Whiskey" and "Dixie Die-way"), and cranked their machines for a week or so, drifting nonstop, until that crust of war was burned out of them. It was as if the first measure for peace and quiet was an absolute blur of mileage, and an exquisite, enchanted exhaustion (something like the *click* Tennessee Williams said you've got coming to you when you're slow-death, heavy-drinking with stern and steady dedication).

Those of us with Vietnam orders looked at the tankers as at a reflection.

Vietnam was going to be awful.

There wasn't any rousing martial air on that plane out of Oakland, not even the mindless, eager expression of war's most puerile sentimentalities and sarcasms; we were private soldiers, lavishly expendable. Rather, the cabin was leaden with the first cloying tinctures, the first raw inklings of a journey that would commence with the fierce grace of banal fortitude (everybody humps their own), move through convulsions of dumb, lightning terror (a deep, strangling clench of the anus simultaneous with the instant intake of an entire breath even as you deliberately struggle to work the machine in your hands; "This is the last of earth"), and end as a psychotic nightmare so removed from the human race that soldiers of wars past have called it another species of man—"We *are* the war" a German soldier once said of the twentieth century's first great world war.

No wonder I cannot recall anything of that plane ride.

And eight or ten months later you took that sullen and exasperated, workaday rage with you on R&R to cities like Tokyo and Manila, Sydney and Honolulu, and other such places, but especially Bangkok; a genuine occasion of quiet, alien ease. Straightaway you hired a bright and pretty young woman to sleep with you, because you wanted to find out if it was still possible to feel good in your body—skip the date, skip the dinner, skip the movie. Most of us discovered to our relief that, yes, we could; but some of us found out we could not (perhaps my brother Philip was one of these). It is not pleasant when you're nineteen to understand that something crucial and precious has gone out of you, even though you cannot say at just that moment what "it" is; not pleasant, I say.

So, if there was any cheer on that plane, it was up front where the lifers sat (the drill of an overseas tour routine), warmly anticipating another stripe, a major's oak leaf, and at least one if not more nice punches on the ticket; and for the guys with rear-area jobs (housecats, we called them)* an easy life of pleasant brutality, as the saying goes.

Larry and I sat a good hour and more, killing time, not talking much, and when our Thai Airlines flight came up on the board, destination Hanoi, we headed down to a small out-of-the-way cluster of gates.

Two planes were scheduled to leave almost simultaneously; a flight to Phnom Penh and our flight to Hanoi. Sitting toward the back of the waiting room were half a dozen or so soldiers of the Yugoslav army sprawled across their row chairs in that way of all soldiers. The sergeant, a thick brushy mustache hiding

---

* The rear-area guys were also called REMFs, Rear-Area Mother Fuckers, and the guys "in the rear with the beer."

his mouth, told us they were on their way to join the UN peace-keeping forces in Phnom Penh, and, with one thing and another, what the hell was Cambodia to them while their own country was going up in flames.

What *can* you say?

Well, Cambodia had to have been the saddest place on the face of the earth; and the old Yugoslavia was on the verge of becoming the next saddest place on the face of the earth.

In ages past the great Khmer Empire was something to behold, and by the twelfth century the Khmer emperors paid tribute to no one from the South China Sea to the Indian Ocean and the banks of the Irrawaddy River. The Khmer built an elaborate system of reservoirs and canals (with corvée labor), which produced four rice harvests per year. Each emperor in his turn built a temple palace at Angkor; the last, the grandest and most memorable being Angkor Wat. It is said that the temples of Angkor constitute the largest religious complex in the world, certainly one of the grandest ever built. By the fifteenth century the Khmer Empire had petered out and all but disappeared, and when French explorers traveled up the Mekong in the middle of the nineteenth century, looking still perhaps for the Northwest Passage, Angkor Wat had for generations been a fantastic rumor (not unlike the sophisticated and elaborate cities of our own Americas).

And during most of the war in Vietnam, Cambodia was distinctly neutral; on paper, anyway.

When President Nixon began the secret (as well as public) bombing of Cambodia, it was said that the men who plotted the B-52 carpet-bombing air strikes could not "box" a target without including at least one town, one village, one farmers' hamlet. Put another way, Richard Nixon's (and Henry Kissinger's) strategic bombing of Cambodia was the vicious, cynical indifference of madmen (a casual, inexcusable racism, really); and if nothing else about our war in Southeast Asia was patently evil, the extravagant bombing of Cambodia has to be the very dictionary definition. There have been some cold motherfuckers in the Oval Office, but Richard Nixon may well be the coldest motherfucker of them all.

Take it all around, Cambodia was a sad place; an aura of the heaviest grief shrouded it. And as remarkable as Angkor Wat may be, I for one had no interest whatsoever to see it for myself.

As close to the border as my battalion operated, in the area between the Fish Hook and Parrot's Beak north of Tay Ninh and the Black Virgin Mountain, my only encounter with Cambodians was in January and February of 1968. The battalion worked briefly with a bunch of Cambodie mercenaries (as we called them; diminutive, if serious, thugs) who were paid, it was said, by the piece. If this had been any of our Great Plains Indian wars, these guys would have been paid by the scalp; counting coup brought them no salt. At night the Cambodies stood around their bonfire, staring down into that rip-roaring glow; high out of their minds on down-home kitchen garden Cambodian marijuana, leaning on WWII-vintage M-1 Garands—rifles as long as they were tall—and packing pliers for the odd gold tooth that might come their way. Out in the woods, when the shooting began, the first thing they did was pop the 24k gold Buddhas hanging around their necks into their mouths so He would be as close to their hearts as possible if they were killed. And in a firefight they were utterly ruthless;

not brave or courageous or heroic, mind you, simply and plainly ruthless.

The spring of 1975, immediately after the war, if in Vietnam there was no out-and-out bloodbath, there certainly was in Cambodia. It has to be one of the most aggressive genocide holocausts of the twentieth century, which was witness to many world-record racial massacres. It is estimated that the Khmer Rouge (communist guerrillas under the leadership of Pol Pot, backed to the hilt by the Chinese) murdered upwards of 2 million people; this out of a population of approximately 7 million. Men, women, and children were summarily executed, murdered by torture, worked to death in labor camps. The young, teenage Khmer Rouge troops (once referred to as the shadow side of Peter Pan's Lost Boys) killed people who wore glasses on the assumption that they were intellectuals. The blanched skulls of the victims were to be seen piled around impromptu shrines by the hundreds, by the thousands, by the tens of thousands (like bleached market-melons), by the tens of tens of thousands; a spectacle all the more grisly and poignant because it is a firm Buddhist belief that your soul will not rest if your head is separated from your body. In the late 1980s, Larry Rottmann drove and walked up-country around Angkor Wat and saw vast stretches of weed-grown, deserted farms and ghost-town villages; not a person for miles, days. An unsettling, disturbing image in a part of the world where it is otherwise possible to stand in the middle of nowhere and *still* be within sight of several hundred people.

The Phnom Penh flight was called. The Yugoslav soldiers, more glum than ever, boarded the bus that drove them across the tarmac to their plane. Larry and I wished them well.

When the bus came back, our flight was called.

*   *   *

The first time I flew from Bangkok to Hanoi in 1990 was an occasion for the avoidance of deep reflection. There I was, going finally to see a place that had always been forbidden to me, even to my imagination. I had no idea what I would find, but I was intensely resolved to take it as it came.

Just the year before I had had a similar experience, and had learned a valuable lesson.

A group of Vietnam veterans (and me among them) traveled to the old Soviet Union to meet and talk with veterans of their war in Afghanistan; this was the Afghan war in which Americans supported the warlord clans, including the Taliban, with money, advisers, and weapons, and which, after ten years, was to end as badly for the Soviets as the Vietnam War ended for us. We flew from Stockholm to Moscow in a light possible only in those high latitudes of the approaching winter solstice. Below us stretched the distinct spectacle of an immense cloud cover as spacious as any desert I had ever seen, as precisely flat as a well-groomed billiards table, and tinged with an intense, crimson sunset as thin as a stretch of waxed knitting yarn. We canted down through what must have been ten thousand feet of solid overcast and emerged to a dismal, late-afternoon darkness with the lights of Moscow all around, and you could plainly see that here was something much more grim than terrible (not even Minneapolis after a deepwinter blizzard was that grotesque). We touched down and taxied off to the terminal past row after row of Aeroflot jetliners parked along the fringe runway, with a small floodlight illuminating the bright red hammer-and-sickle symbol on the tail of each plane.

And it all came flooding back to me: a Cold War childhood

of duck-and-cover drills; the persistent propaganda that an international conspiracy threatened the whole civilized God-fearing Christian world; of provocateurs (with superhuman cunning and the Devil's guile) conniving to overthrow the government (including the very building where I went to school); the nightmare fright and whipped-up hatred of "the Russians" who were going to make slaves of the ones they didn't kill, cook, and eat right then and there.

We were met by our Russian hosts and given the traditional Russian welcome of bread and salt right then and there; nice touch, that. They called themselves *afghantsi* with much the same wry, ironic élan as we now called ourselves "grunts"—these guys in their early and middle twenties, superbly robust, and young enough to be my sons (I was abruptly struck just then at having lived long enough to *have* grown sons).

Their war in Afghanistan was a two-year tour for armored cavalry troops and airborne infantry (tanks and tracks on the road, and airborne assault by helicopter); no R&R, and, of course, no cathouses in Kabul. The official Soviet story was that the troops were there to build schools and hospitals, to plant trees; "nation-building" is the modern sarcasm. The Soviet army depended on conscripts, except for the sons of guys who had enough pull to get them exemptions of one kind or another (the same as here during the years of the Vietnam draft), and this was, make no mistake, a source of friction between the haves and have-nots (the same as here). The bodies of those men killed were shipped home in sealed zinc coffins (hence the term "zinky boys"), the families were *not* allowed to see the corpses, and the funerals were held at night to discourage crowds (and the talk); and when the men finished their combat tours and rotated home, they were forbidden to speak of the war and when they did were disbelieved and hooted down. The *afghantsi* spoke of these things with a bitterness that did not require an interpreter; they spoke of the Afghan rebels, the mujahedin (supported by the United States), with undisguised hatred (often referring to them as "muckheads"). It took us a week to convince the young *afghantsi* that the infantry soldiers in Vietnam were generally conscript draftees just like them, and not mercenaries (as they had been told), and paid an ordinary soldier's wage.* It took them a week to finally declare (with a sly grunt's pride) that, yes, they "did" drugs; that, in fact, in every village market you could buy Sony boom boxes, fresh vegetables, and good old hashish by the plug. At Moscow University Law School the young ex-soldiers in coats and ties talked of the changes coming hard and fast (Gorbachev's perestroika), and if the changes didn't go their way, well, they could get plenty of explosives, and the AKs and RPGs were no problem—the sort of blunt talk from ordinary soldiers that I hadn't heard for many, many years. The Soviet communist system was coming undone and, according to these guys, the end wouldn't come a moment too soon for their taste; it was clear they were fed up, and *that* with a capital F.

After two weeks of simple talk, many late-night sessions trading war stories, flipping through many, many albums of photographs, and passing around many bottles of ice-cold vodka, everyone in the saloon, from Missoula in Montana to Alma-Ata in Kazakhstan, understood that for ordinary soldiers the only difference between Vietnam and Afghanistan was that Vietnam was green and Afghanistan was brown.

I did the math once and my pay in Vietnam worked out to something like thirteen cents an hour, which included the additional perks of overseas and combat pay.

Some months later when that war finally *did* end and the last Soviet troops drove away, a general was asked if the Soviet Union had made a mistake. He hemmed and hawed for a moment, and then said, "There is an old Russian proverb, 'A tailor should measure his cloth seven times before he cuts it.' "

It was at the Moscow airport, oddly enough, that I saw the first Vietnamese outside the United States since the war. They were volunteer laborers and looked cold—none of those twenty or thirty kids were dressed for the weather. I asked one guy where he was bound. The coal mines of Armenia, he said. Well, good luck to you, young man; not a week later we heard that there had been a mine disaster in that neck of the woods, and my first thought was for those Vietnamese kids. A year later I asked a Hanoi poet-buddy of mine about this matter of contract labor (of which I was very skeptical), and he said that those jobs were highly sought after; that his brother had worked what could only have been some horrible job shoveling shit in an East German factory, but two years later arrived home with enough money in his poke to buy a 50cc Honda scooter; a status symbol of singular rarity in Hanoi in those years.

I say, on that first writers' trip to Vietnam in 1990, we took a less than half full Vietnam Airlines flight from Bangkok (the first couple of rows piled with luggage), circled around Cambodia, giving that sorrowful place a wide berth; flying through the panoramic haze of the late-season monsoon. There was "weather" everywhere; the very air *looked* blazing hot. Perhaps this is what they mean when they say the sky is too hot to be blue. As we sailed over the mountains west of Hanoi, I was told to look for the swaths of B-52 bomb craters, and sure enough as we descended through the smoky heat of the Red River Delta there were plenty; then more and more the closer we got to touching down. It was as if a scarf of a million bombs had been blind-stitched into the topography. We also glided over the single-track railroad main line that connects Hanoi with Lang Son on the Chinese border; from that altitude the rails were as thin as a pencil line and precise as a draftsman's French curve—like the clean whittled curl of a cedar shaving as long as your arm.

We landed behind a well-used Soviet Aeroflot wide-body arriving from Moscow, a machine notable for its bulked-up cheesiness and said to be easily converted to a heavy bomber (the exit stairs take you through the bomb-bay). We parked some distance from the planeload of fat and pink, sweating Russian tourists come to eat, shop, and generally bask in the tropical resort of a "socialist workers' paradise." A double-axle flatbed truck backed up to our plane's cargo door. The truck vaguely resembled a 1930s International Harvester or a Chevy of like vintage; some banged-up old beater with the look of a veteran of many a round-tripper on the Ho Chi Minh Trail. If you asked someone what kind of a truck it was, the guy would probably look at you with considerable bafflement and say something like, "Well, it's a truck. Made in China at Big Truck Factory #44. You know, a *truck*." A bunch of Vietnamese day laborers climbed through the cargo hatch and started unloading baggage and boxes of "stuff" onto the flatbed—it reminded me of bucking hay. We passengers began making our way across the tarmac, ambling along, talking and gawking, walking through the hot, roundly heavy smell of Vietnamese earth.

If the Bangkok airport had a golf course, of all things, on

the grounds—fairways, sand traps, and greens scattered among the fringe taxiways—Hanoi had vegetable gardens; rows of cross-hatched bamboo pea fences, and melon vines; kitchen-garden rows of corn and cane and salad greens. Beyond the airport security fence a vast stretch of lotus bogs; yonder, rice paddies with hamlet groves secure behind hedgerows of bamboo thickets (as elderly and bluntly fortresslike as *le bocage normand*, the famous hedgerows of Normandy in France); and way back of that, more bamboo hedges, oxcart trails, and footpaths, then other farms, more hamlets, rice and bamboo. And way past *that*, Hanoi, city of the rising dragon, established as the capital of the Ly dynasty in the eleventh century.

We walked into the terminal, and, as with many other buildings put up since the American War, it had the look of being thrown up in shoddy haste with Soviet aid. If form was function, here was a horrible place to work (unless you were glad for the job).*

Everyone's luggage was delivered to a little bitty room on an industrial-strength baggage carousel of cranky, hard-rubber slabs. What little light there was came through the doorways, high windows, and skylights. The Russians, as a group unhealthy-looking—not so pale as ill, rumpled and exhausted and thoroughly wasted on jet lag—waited for their bags with a luggard exasperation that spoke of a lifetime of waiting in line; grumpy-looking, dour, singularly unhappy (as if the brochures never said anything about the *warm* weather; as if

---

* *In 2002 a brand-spanking-new, state-of-the-art airport terminal was opened, as handsome, even dazzling, as any you may have seen lately; this accompanied by an equally modern four-lane interstate-looking highway that will take you into town lickity-split. The old warehouse-looking terminal is long gone, and good riddance.*

they had been *sent* on vacation), all those people wanted to do was get to their rooms at the Cuban-built guesthouse compound on West Lake, crank up the air conditioning, peel off their shirts, pour themselves the first of many, many tall glasses of store-bought vodka, and get dead drunk as quickly as possible. We would see them several days later, delivered en masse downtown in tour buses, lumbering among the shops of Old Hanoi, picking-at and fingering the merchandise, and all but deliberately ignoring the Vietnamese.

The Vietnamese didn't much like the Russians, but put up with them with a patience that speaks volumes of their mature wisdom. I asked my Hanoi poet-buddy about that, and he said that the Russians are just too cheap; in the market, for instance, well, everyone banters and gabs, turning things this way and that—you know—haggling prices. The Russians don't shop and bargain, they argue, he said, and with a sour expression on his face.

Our Writers Association hosts came straight past the air-port terminal police line, on through the building, and greeted us with a suddenness of lively hellos that surprised everyone in the room. There was a complicated round of introductions, and the younger Vietnamese writers fetched our luggage. We were hustled through a rhetorical formality at customs (the cops and customs guys schmoozing with several of the writers, who might as well have been rock stars), then loaded into Honda vans and driven to town.

That was the first trip.

This time, Larry and I sat on the plane, ate a very quick lunch, and read our books.

3

Funky, Beautiful Hanoi

It was May, and hot in Hanoi.

Larry and I arrived at the Noi Bai International Airport across the Red River from town, drifting down through hazy skies over the countryside, and touched down—easy, easy.

This time there were many fewer bomb craters. In fact, the craters were almost gone, no doubt filled in by many hands with many ditch spades and many baskets of dirt hauled two at a time on many shoulder poles the same way the Vietnamese had shoved and dragged their artillery to Dien Bien Phu and later built the Ho Chi Minh Trail and the Cu Chi tunnels: the same way they had rebuilt the railroad from Hanoi to Ho Chi Minh City, and the same way they had rebuilt their cities, towns, and hamlets (brick by brick); the same way, in fact, the

Vietnamese had settled and built the whole region from the Red River Delta to the Mekong (River of the Nine Dragons), the same way the Vietnamese had been digging out from under the aftermath of two wars (with the Second World War thrown in for "lagniappe"). These guys have nothing if not patient fortitude.

We were met by the novelist Le Luu and the diminutive and indomitable Ms. Dao Kim Hoa, translator and speaker of five languages, organizational powerhouse, and senior administrator of the Writers Association—our hosts for the trip.

Le Luu, one of those guys who always has a smile on his face, though he doesn't have a good tooth in his head, is the author of fourteen books. He grew up in a village of small-hold farmers along the Red River east of Hanoi, and talked of the annual spring flood in much the same way that Samuel Clemens and William Faulkner spoke of the high-water Mississippi. When he was a young man, in his turn he joined the army in 1964, "Went down the trail" (as the Vietnamese say), rose through the ranks to colonel, and returned north in 1975 with the rest. His writing has always been popular, but to give you an idea of the writer's life in Vietnam, when *A Time Far Past* was published in 1986, Le Luu made just about enough money to buy himself and his family a 50cc Honda scooter, exceedingly rare in Hanoi in those days.

Kim Hoa scooped up our passports and papers and *took* us through customs, running a gauntlet of fellow travelers and blasé customs cops; our several pieces of paper glanced at and stamped with an official chop and a whirligig signature; our luggage waved through with a slow-motion wave; the rooms close and the ambient light dim; everyone in the place greeting

Le Luu with surprise and abundant compliments, small-talking with him as if they'd known him all their lives.

In those first moments, from the opening of the airplane door to the warm, physical greetings of Kim Hoa and Le Luu to the hauling of the luggage through the shoulder-to-shoulder curb-side crowd of freelance "cabbies," parties greeting travelers, and the downright curious come to the airport simply to gawk at the rush-hour parade of big-nosed, round-eyed foreigners—in those first moments that strange oppression of the unfamiliar heat and humidity squeezes you down, and you have to ask yourself what*ever* did the French, pouring sweat in their white flannel tropicals and those nifty white pith helmets and dragging behind them every last blessing of European Civilization behind them, *what*ever did they see in this place? (It *never* gets this hot and humid in Chicago.) The weather reminded me of Samuel Clemens' crack about the two kinds of weather in India; the one kind that will melt a brass doorknob, and the milder sort that will only make it mushy.

This is discouraging, you say to yourself, because it's going to be like this morning, noon, and night for the entire trip.

But by the time we settled in the small and cranky four-door Writers Association car (of Soviet manufacture: it smoked; it growled; the seats and pedals were worn thin; there seemed to be a lazy swerve in the steering), pulled out of the parking lot, and were on our way down the road toward town, that discouraging feeling of heat and oppression passed.

It passed with the warm breeze blowing in all the windows (ruffling everyone's hair), the shared smokes, the talk of family and who among the writers had a new book out. So-and-so was in Saigon working for the Health Ministry translating articles about AIDS (much to everyone's surprise, there had already been a handful of deaths from that disease); so-and-so was at a conference in Helsinki; and so-and-so had a new baby, daughter number three. We passed the gauntlet of industrial billboards (hawking machine tools and office machines and heavy-duty construction equipment), the ma and pa convenient-style Petrolimex gas stations, the stacks of bricks, bags of cement, and piles of sand at house construction sites, through roadside villages with their vegetable markets littered with kitchen-garden trash, disabled trucks with one axle up on homemade blocks, the storefront bike repair shops, and panorama of rice paddies and the odd bomb-crater-turned-carp-pond not a mile from the airport.

And, funny thing, the air changed, and it felt (oddly; oddly) *good* to be back in Hanoi—a thing I never thought I'd say in *this* life.

There are those cities that invite affection; those places on the earth where a body understands with absolute certainty that an affinity exists between you and it, and the connection is keenly felt in both directions. I have lived in Chicago all my life; in fact, I live about two miles from where I was born, and, as it happens, two miles from the house where my father was born (my family has lived in the Edgewater neighborhood for one hundred years and more); the only time I was away from the city were the two years I spent in the Army. My wife and I have moved three times in thirty-six years; we can see our old apartment around the corner and down the street from our kitchen; it is almost un-American to stay in one place. I am rooted here, as nowhere else; Chicago, the city, the river, the lake, the prairie, is my "place." But, then, so are Seattle and Oakland and Boston. Dublin, Galway, and London. And here is Hanoi, the city of the rising dragon for one thousand years, give or take; funky, bustling, positively elderly; the groove of Hanoi life is scratched deep into the Vietnamese character and goes back a good long reach.

As I have said, for more than half my life Hanoi was a forbidden place—forbidden even to my imagination. Now I have good friends here, perhaps the best of friends, because I share with these folks something I share with very few in the States. If my war-year still arouses deep and unmistakable memories (all vivid, none pleasant), and the writers' trip of 1990 was a welcome discovery and revelation, then all these subsequent visits have been warmly anticipated and richly enjoyed. This is simply a beautiful country; the food is great, an amalgam of Chinese and French cooking (two cultures that know their way around a kitchen). The women are beautiful, whatever their age. And for a train buff like me, riding the trains is all kick. Let me be clear. These sojourns are not "guilt trips." And I don't go to Vietnam to "heal" myself with one of those good, cleansing New Age crying jags. No, "healing" is too dicey a business to be settled with a couple weeks' vacation. And there's nothing about "reconciliation" in these visits; I was a soldier once, as was just about every Vietnamese of a certain age I have ever met; we know what that means, and leave it at that. I've come to see that the Vietnamese have a deep (not to say historical) sense of melancholy, and are more than empathetic with Americans returned to settle a grief that will not sit. If there are Americans (veterans and kin) who travel to Vietnam on unfinished business that has nagged them all their adult lives, then the Vietnamese also travel—south, to look for sons and brothers, husbands and fathers gone missing since the war by the tens of tens of thousands. So the Vietnamese of the north have their own bereavements and griefs to hone off. But we will come to that part of the story by and by.*

We drove south toward the city through countryside, the scrubby, gruff-looking industrial suburbs, past the railway shops in Gia Lam and the Red River levee, and crossed the low-water mudflats of the Red River (it reminded me of the Mississippi). Just downriver, on the Long Bien Bridge, a small Soviet diesel was pulling a dozen hard-seat day coaches toward Haiphong.

We came off the newly built toll bridge, a spiffy modern piece of work, and over the dike and down to the street, and the city was right there, spread out before us. Block after block of shops open to the street with apartments above. Shade trees, rough canvas awnings, round-tile roofs streaked with dark, tropical-city dirt, and huge potted bougainvillea cascading over the second-floor porch railings above the shops; bonsai and flowers everywhere. Trucks and buses, 50cc Honda scooters and three-wheeled pedal-powered cyclos, and many, many bicycles; street vendors selling lottery tickets, amber-colored petrol in two-liter jars, fresh fruit and cut flowers, and cigarettes by the pack or the smoke; and foot traffic—women haul-

---

* I have not heard anyone speak of the French returning to Vietnam in the same spirit that Americans travel there; the Vietnamese war against the French was a different thing altogether. I suppose that by 1954 the French were eager to be shuck of Indochina.

ing baskets of produce and whatnot on shoulder poles—and folks loafing in the shade of red-blooming flame trees, as the Vietnamese call them; a harbinger of summer. The air was hot and dry, smoky with the dust of many tires and many feet.

I have three vivid memories of my first visit to Hanoi.

One of our very first "tourist" stops was to the Temple of Literature, the Van Mieu, Vietnam's first university, established almost a thousand years ago by the emperor Ly Thanh Tong (after the Vietnamese got rid of the Chinese that time). It was here that the lucky, lucky few who passed the local and district Confucian-style exams would come and study for the final that would establish them as high court mandarins (sweet-deal, high-government, life-tenure jobs of a very special kind; a man could be middle-aged before he made the grade). In theory, anyone could take the tests, just as, in theory, any native-born American at least thirty-five years of age can become President of the United States. In front of the pagoda-like building where classes were held is a garden of flowers, flowering trees and bonsai, and a broad carp pond—the Well of the Quiet Heart. Set round about were eighty-two man-high stelae on the backs of tortoises (the animal said to have brought knowledge into the world, and here the size of tractor tires) declaring the names of the graduates.* As we were walking back to the van, ahead of us was a guy sitting on a low stool at the curb

* I read somewhere that there are fewer than a thousand graduates of the Van Mieu, which ceased giving the grueling tests, modeled after the traditional Chinese exam only in the early part of the twentieth century, when the French began closing down many a Vietnamese school. Pretty stiff competition, ask me.

selling snakes out of a wire cage. When he saw us, he reached into his basket and grabbed a cobra, set it on the walk, and it began to crawl toward Bruce Weigl, who is not a small man. When the snake had slithered almost out of the guy's reach and he judged that we were within picture-taking distance, he slapped it on the tail and the snake rose knee-high and fanned out its hood, hissing, so that Weigl (just *that* quick!) could stop a moment, crouch down, and take a picture. The guy laughed and Bruce laughed, and we all laughed—*Gotcha*, just like the playground basketball grab-ass rule (apparently the world over), "No harm, no foul."

Then later we were strolling through the market neighborhood of the oldest part of the city. We had to be the only people on the street; it was hot. We came up to an old woman sitting on the curb under what little shade was to be had under a tree with a winnowing basket of lotus flowers picked fresh just after dawn that morning (juice gathered at the stalk-end) arrayed before her. And as we walked by, she absentmindedly picked up one of the flowers and began brushing the sparkling white blossom over her face, so that the pollen clung to her eyebrows and lashes, and dusted her cheeks. It reminded you of Deep South church matrons fanning themselves of a Sunday.

Later that evening (it had to be close to midnight) we were driving back from an evening of water puppet theater, cruising the backstreet neighborhood cabdriver shortcuts. The city was plenty dark; there was hardly any light in the streets, and what little there was spilled out from block after block of little hole-in-the-wall shops and cafés; from hand-sized kerosene lamps and smaller perfume-bottle-looking jars. The streets and storefronts, the promenades and bits of park, the rows of bikes and crowds gathered in cafés (watching World Cup soccer from Italy with an intensity that Americans save for the World Series or the Super Bowl), the strollers and the folks out late to catch the cool—all that Hanoi nightlife seemed afloat in an aura of butterscotch air. You knew it was an optical illusion of one kind or another, but still, with the late hour and that cast of little light, the pavement in front of us that seemed to simply melt into the darkness ahead of us, loafers and strollers and bicycles and cyclos seemed to float in air; a singular memory, I'd say.

But then, we were *always* bumping into the inexplicable; young cops almost embarrassed to be hassling you; or the chickies who would squirt past you in the Ho Chi Minh City traffic, shouting, "Follow me!"; or the restaurateur in Hue who stood at the gate of his establishment, thanked us for coming, bid us a good evening, and reminded us to tell our friends about his place, "I've been reviewed in the *New York Times!*" and you shake his hand with more warmth than ever, replying, "Me too!"; or bumping into a Hanoi concert pianist, the guy just back from Europe, and *that* quick we knew everything there was to know about the other, and congratulated ourselves with much warmth that we had lived long enough to write and play the piano—well, sir, ain't life sweet!; or the ex-infantry colonel sidewalk barber, who whiskey-whispered in my ear that, "We were *all* crazy" as he finished off my trim with his straight razor, working that thing around the back of my neck (that scrappy *zing* sound, accompanied by the "haircut shiver" down the back).

The Vietnam Writers Association arranged for Larry and me to stay in a family-owned guest house south of downtown on Nguyen Du Street (pronounced N'win D'zu, after the nineteenth-century poet*). It had the look and feel of a bed-and-breakfast place without the sitting room and concierge; something like a five-room SRO.

Our "house" was across the way from a large lagoon that covered several square blocks, called Thien Quang, or Light of Buddha Lake, where on the weekends you could rent a swan paddleboat and cruise around in the sweltering sun—which, of course, we never did. Every morning, first thing, a couple of neighborhood guys would hand-paddle a small boat around the lake, netting fish, which they sold (still kicking in the net) right then and there to sidewalk passersby. And, since this was a neighborhood with many smaller hotels and guest houses for foreigners, the hookers (an embarrassing inevitability to my Hanoi friends in these weird economic times of *doi moi*) would hang out around the lake across the way at the corner of Tran Nhan Tong and Quang Tung Streets. And across from the lake, yonder was Lenin Park (built after the French left in 1954 atop a wetland swamp that was little more than a city dump).

Our guest house was a converted French colonial residence (of the middle management kind, as was every house in the neighborhood). The family who owned the place lived on the first floor, and there were five rooms upstairs with the maids' rooms, laundry rooms, etc., toilet, and shower. I don't think we ever met the owner, but Larry has a wonderful picture of the day

---

* *Nguyen Du's claim to fame is a 3,254-line epic poem The Tale of Kieu; a story of love and family duty and karma; considered by the Vietnamese to be the greatest poem of them all. It is something of a parlor game to recite stretches of the poem, or to let the book flop open, pick a line with your eyes closed, and interpret (fortune-teller-fashion) the line for yourself.*

man sitting in the shade of the carport with a Hong Kong oscillating floor-fan roaring in the doorway next to him. In my
room, the back windows (framed with thickly painted, louvered French shutters) overlooked the courtyard with its tiled
carp pond filled with flowering lilies and hand-sized carp, the
whole space covered with netting for the bright-blooming cucumber vines; the cukes hanging down like bulbs of wax. All the
houses on the block, and indeed in the whole neighborhood,
were fronted with high brick and stucco walls with iron gates of
one kind or another closing off the narrow driveways. (To be a
French colonial bureaucrat in the old days, and there were at
one time forty-five thousand of them, must have been cushy
work judging by how they lived; servant labor dirt cheap and
"natives" glad to get the work, everybody waiting on you hand
and foot.) The room was furnished with a large double bed
(festooned with a mosquito net), a Soviet air conditioner (the
dials marked off with Cyrillic instructions and Roman numerals), a highboy wardrobe, and a couple chairs. On the table between the chairs was a glass carafe of water, glasses, a teapot
and little bitty cups, and a tin of tea; on the floor near the chairs
was a large thermos of very hot water (a common room accessory everywhere in Asia) and a plastic spittoon-looking thing
where you disposed of the tea turned cold. And all the rooms in
all the hotels and guest houses in all the cities we visited were
pretty much the same; only our hotel in Ho Chi Minh City had
a television.

Since everyone walked or rode bicycles, heavily muffled
50cc Honda motorbikes, and scooters, street noise was almost
nonexistent—a thing so pleasant I cannot begin to tell you. Automobiles were rare and private cars unheard-of. What truck

traffic there was stayed to the boulevards, unless the guy was trying to make time with a shortcut.

And even though the traffic moves just fast enough to keep the slack out of a bike chain (it seems), Hanoi is small enough (about 1 million souls) so that we could get anywhere we wanted to go in a half an hour of easy pedaling. Of course, it helped that we were living in the middle of the city.

After we got settled, the first thing we did was check out the neighborhood for the local restaurants—first things first.

In what traveling I've done, I've come upon a couple hard-and-fast rules: "Never eat in the hotel where you're staying and always drink the local beer." This for a couple reasons. In a hotel you will see only your fellow travelers and the staff, and the menu is always the same. And, too, there is no better, faster way to get to know a culture or a people than to eat their food, prepared as they would eat it—a body can always get a hamburger and an order of French fries, or a deep-dish pizza, when you get home.

The seasoned traveler will not fail to notice that the Vietnamese, like the Chinese, will pretty much eat anything that trots by the cabin, as the saying goes. Chicken and beef, seafood and pork are staple meats, but you'll also see an abundance of such generic menu-exotica as "Bird," "Snake," and "Dog." So, if the "mystery" meat looks a little too much like roadkill for your taste, eat a little to be polite and let the rest go (it will *not* go to waste). If something unfamiliar looks unpalatable, it's best not to ask what *it* is unless you really want to know; it might be a species of traditional, American household pet, or a garden or alley pest. And in Vietnam, as in all Asian countries, rice is served with every meal. There is an old saying that, indeed, "Rice is the wife of Vietnam"; a nice way to put it.

You are never very far from the ocean or a river or a farmyard carp pond, so there is a lot of seafood. At the Non Nouc Hotel at China Beach one evening I ordered boiled shrimp and was presented with four of the critters, each the size of a banana. And I understand that before the American War, Nha Trang was famous for its lobster (a local breed, like the Dungeness crab of the Olympic Peninsula), and that during our war the Nha Trang lobstermen quickly fished it out (with dynamite, among other things, destroying the habitat); but I'd also heard that Nha Trang lobster were on the comeback.

On my first sojourn to Vietnam in 1990 with the writers, we couldn't help but notice dogs everywhere in the countryside; the scruffiest, most mongrel-looking, most pathetic hangdog-looking dogs I had ever seen—"composites" is perhaps the polite term. Then again, there also seemed to be larger, well-cared-for, important, and confident-looking dogs. Now, at some of the meals we were served meat that distinctly was not any recognizable cut of meat I had ever seen. We asked and were told it was dog. Well . . . , okay. It's a little stringy, and there didn't seem to be much on the hock, but well, fine, dog. Eat up (don't make a big deal of it). Finally someone asked Le Luu how a body could tell the difference between the food dogs and the pet dogs. Le Luu grinned big, laughed big with his whole body as if he'd been waiting for someone to ask, and answered, "The *dogs* know." And by the look of them the dogs knew plenty.

In fact, the difference seemed to be that the food dogs stayed outside with the other livestock and got fed (mostly table scraps), while the pet dogs got the run of the house, were petted and taken care of, and got fed (also table scraps; nothing goes to waste, you understand).

There was a good restaurant at 59 Nguyen Du Street, just down the block, which we came to call the Sourpuss Café, because that first morning the waitresses didn't seem glad to see us at all—perhaps because we showed up long after the morning breakfast rush (8:00 A.M.)—good God, more work. When we came back morning after morning, engaging them in what little conversation can be had with limited vocabularies (slowly enunciated English, some grammar-school Vietnamese, a smattering of pigeon-French and GI-German as well as outright pointing and creative body English), the waitresses' attitudes changed, but the name—Sourpuss Café—stuck. We'd walk into the small room of stucco-covered walls, take our pick of tables, turn on the floor-fans, look at the menus (in Vietnamese and French), and try to order—usually omelets, French bread, fresh-picked in-season fruit, and coffee. The waitress would take the order back into the kitchen; someone would get some money from the cash box, hustle over to the market down the street and around the corner, and buy the fixings. The eggs were so fresh that some still had the farmyard dirt and straw on them.

The kitchen women were generally delighted with us because we would flirt and carry on and play our own tapes (my homemade put-togethers of Chicago blues, rock 'n' roll classics, and Larry's Ozark mandolin, banjo, and fiddle "jazz") on their large semiportable Sony boom box with bad bass speakers—of a kind and size found in any self-respecting Vietnamese café. And too, because one morning Larry went into the kitchen to *show* them how to cook an omelet—"Missouri Ozark" style.

Before then, if we asked for one, they'd bring us plates of duck eggs fried to death on both sides (the pullet eggs are too small); not what anyone—especially the French—would call an omelet with a straight face.

Larry let it be known that what they had been serving us were not omelets at all. He went back into the kitchen and called for the iron skillet, eggs, some green onions, cilantro—whatever was at hand. They looked in the old, round-top refrigerator and saw that the eggs were "finished," as they called it. Someone was sent to the market to fetch half a dozen. Not a moment later back she came with a basketful of eggs and other fixings. Larry grabbed a large bowl, whipped up the eggs, chopped up some scallions (with one of those large, flat Chinese-style cleavers), diced a couple tomatoes, some asparagus tips, hot green peppers, and sprigs of cilantro, added them, sprinkled in several pinches of salt and a long pour of condensed sweet milk, then turned to the old gas stove and greased up the skillet with lard. In no time at all he had rustled up two tolerably good-looking omelets, working the skillet and spatula with the grand and showy flourishes of a Toddle House short-order cook while the kitchen women and several late-arriving passersby watched with sputtering astonishment that soon transformed itself into gleeful amusement and high humor. (It wouldn't surprise me to learn that his recipe was added

to the menu—stranger things have happened—*omelette à
l'Ozark américaine.*)

These Americans!

In the other direction, the guest house was two doors from
a brand-new place called—I am not making this up—The Boss
Hotel, complete with cab line, white canvas awning out to the
sidewalk, uniformed doormen, Italian chase lights draped all
over the four-story façade, and a karaoke/disco VIP Lounge—
"Where Every Guest Is a V.I.P."—where you could dance with a
young lady (for a price) and arrange for a massage later (for a
considerably steeper price). You'd walk by the place at night,
tiny white lights flashing and twinkling, and think of the river
district of crummy strip joints in Louisville, the Susie Wong
Bar in Bangkok (where the young women wore large numbered
badges not unlike the floor traders at the Commodities Exchange at Chicago's Board of Trade), and the frumpy and elegant joints across the street from the Tokyo Enlisted Men's
Club, where the R&R hookers walked a track in polished, edgy
elegance. I know this sounds callous, but a young woman working in the sex trade in Hanoi can make as much in a week as her
father makes in a year; it is, I suppose, the downside of a
daughter's sense of filial responsibility. It worked the same
with an emperor's concubines. A family would almost literally
serve up a daughter to become one of the emperor's women,
and the family would accrue status for that—not to say social
clout. Once a young woman stepped over the threshold of the
citadel to join the man's extended family, it was good-bye—
never to be seen or heard from again. And when the old emperor died, she along with all the other women-not-his-wife
would be put aside by the new emperor (who would begin col-

lecting his own cadre of women), and wind up doing other scut work in the palace. For certain, she could not return to her family, and for dead certain could never marry.

On any city block there will be at least three curbside entrepreneurs: a bicycle repairman, a woman selling cigarettes by the pack or the smoke, and a young boy selling lottery tickets—the Vietnamese are great gamblers. Mixed in with these are the fruit vendors, hole-in-the-wall cafés and tea stands, noodle shops, and perhaps a vegetable stand and a billiard parlor. In our neighborhood there were also a couple of local street markets with chickens, pigs, and dogs, more fruit and vegetables, clothing stalls, T-shirt and hat stalls, shoes and kids' clothes stalls, and the Laotian and Australian embassies. Also wandering around were young guys, perhaps fifteen years old or so, selling Vietnamese and foreign-language magazines and newspapers—maybe a dozen titles. They held a piece of cardboard down their left arms, a portable rack, with the magazines and newspapers arranged just so. They'd come up to you, three or four at a time, and, I'm sure in the only English they knew, asked if you'd like to buy. There was never anything in English, so what was the point.

Everyone in Hanoi (and everywhere else) was quite ingenious and artful about finding some sort of hustle or other. Even regular wages-work did not pay enough, so many folks also had a little something on the side, and everyone in the family, the household, pitched in and contributed something. Even if you had few resources and couldn't find anything else to do, you could always get yourself a bicycle pump and sit out on the curb, charging a nickel per tire.

On our way to breakfast we'd say hello to Mr. Pham Hong Khanh, the bicycle repairman who sat under a tree at the curb right outside our gate. Hong Khanh was older, always sitting in a high-backed bentwood chair, dressed in long pants, shirt (buttoned to the collar), jacket, plastic sandals, and black beret (the fashion among older, and therefore French-educated, men). He had a small portable workbench about the size of a milking stool and an old GI ammo can full of tools of mixed vintage, and sat watching the traffic with his legs crossed and his hands in his lap (when he wasn't reading). He sat straight-backed with a most dignified air, made odd by the fact that he held his glasses to his face with strings looped around his ears. When we first encountered each other—me asking him to look at my bike and give the tires a shot of air—I asked him if he spoke English, to which he answered no he did not, but did I speak French, which I did not; at least that was the gist of the conversation. His whole graceful manner, his upright carriage, made me think that this cannot possibly be what Mr. Pham Hong Khanh does for a living. But whatever else he was, he was a good bike mechanic; always cheerful and helpful, greeting me coming and going with a smile and a neighborly nod. And when I showed him pictures of my family, the house in Chicago (one picture taken in summer and one in winter with snow to the very windows), and got him to understand that I was from the U.S. and an ex-soldier, he seemed genuinely glad to make my acquaintance and looked forward to seeing me every morning. (He knew about Chicago, though we're still famous for our gangsters. He well knew of Cu Chi, because of the tunnels, and Tay Ninh because of the Cao Dai Temple, but his face positively lit up when I mentioned Nui Ba Den, the Black Virgin Mountain, because he well knew the story of her soldier-lover and of her suicide.) Afternoons, Hong Khanh was gone, and I assumed that he was working another corner, home taking a nap and watching the grandkids, or sitting at his wages-work job; then one day he didn't show up, and the other curbside entrepreneurs had no idea where he'd come from or where he'd gone. (Wherever you are, Hong Khanh, greetings to you.)

After a moment of checking the weather with Hong Khanh, it was on to the neighborhood noodle shop.

Up Tran Bing Street, just around the corner of Ng Gia Thieu, was the neighborhood curbside *pho* shop. As near as I can understand, almost everyone eats *pho* for breakfast, and these little curbside, housefront enterprises amount to Vietnamese fast food.* It's fancy chicken broth, rice noodles, a meat such as chicken or beef or seafood (or what have you), garnished with French-cut scallion onions, half a handful of Vietnamese basil, and served in a large bowl with a wedge of lime and a jar of hot sauce on the side to taste. *Pho* is inexpensive and filling, more than enough to get you cracking first

---

* Pho, *pronounced "fe" with a short, soft "e," is basically spiced chicken noodle soup, which some people in this world think will cure anything. Vietnamese is a tonal language, like Chinese, and as you might expect the same word can mean several different things, depending on the accent. There are six accents, and children are taught them by associating them with different animals. Ga means chicken, but it also means train station, and "to permit"; it is a form of the third person pronoun, but with a tinge of disrespect and disapproval; it is a term used to describe exaggerated and cloying salesmanship; as well as the paying of a gambling debt with trade or goods. You can get yourself into big trouble if you put the wrong accent on a word, but the cook at the noodle shop understood perfectly well that I was asking for chicken soup, even though I may have said train station.*

thing in the morning; *more* than a welcome change from the cold and dry, sugar-heavy breakfast cereal that most Americans choke down and call *good*.

The family that ran the place no doubt started making the broth very early in the morning. There were half a dozen tables and benches and stools set up picnic-fashion on the sidewalk along the curb, everything low to the ground. And when I arrived in the morning around seven, the place was crowded with ordinary working guys stopping for breakfast on their way to work. At the curb, under an umbrella awning, is the kitchen—a collection of pots and tubs with dad, the cook, standing in the middle. He stands among the large bowls of fresh scallions, diced meat, pots of cooked noodles, a tub of simmering spicy broth, and other tools of the trade. The mother sits among all this minding the cash box. Inside the garden gate a couple of the kids sit on their heels cutting up the chicken and meat; you walk up to the guy (you standing in the street) and say, *"Pho ga."* Pho with chicken. He takes a handful of noodles and drops it in a strainer about the size of a derby hat and drops that in the tub of boiling water set on the roaring charcoal brazier; this to quickly reheat the noodles. Then he takes a large soup bowl that holds the better part of a quart and drops the noodles in and ladles in the broth. Then he scoops up a handful of diced chicken and adds that. Meanwhile, you sit as best you can with your knees well up on a low, rickety bench or stool (everything made of rough-cut mahogany planks and knocked together with whatever nails or pegs can be got in the hardware aisle at the local market), pick out a pair of chopsticks, and an oriental-style, flat-bottomed tin spoon. One of the daughters brings your bowl of *pho* and the French bread. Squeeze yourself some lime juice, squirt in some hot sauce to taste, add some Hanoi basil and bean sprouts, swish and mix everything around with the chopsticks, and help yourself. (If this sounds spicy, and it certainly is, I quickly learned on my travels southward that northern Vietnamese all have the opinion that *pho* in the south is just *too* hot; which I do not agree with—peppery *hot* food; at least you know you're alive.) The Vietnamese eat with both hands; that is, chopsticks in the right hand and that spoon (almost a ladle) in the other, working both at once, and it's plenty noisy; the thought seems to be that if you're not making any noise with your food, you're not enjoying yourself.

You eat and watch the traffic, talk to your fellow diners who just might speak English, and pause to take pictures. The guys sitting around you smile and nod, amused that someone who is definitely not from around there has come to sit among them for breakfast. You are perhaps the subject of breakfast conversation (of which you know nothing because, of course, you can't make head or tail of their whispered Vietnamese).

One guy (wearing a baseball hat with a California tool company logo tilted back on his head) we kept seeing in the morning laughed out loud and rolled around on his stool when he saw us coming. Apparently our sitting down among ordinary working-stiff Vietnamese amused him no end. Finally one morning I asked what was so funny. Who knew, perhaps it was the way we held our chopsticks or looked so funny with our knees up into our chins (Larry is a good bit taller than I, so his knees stick up a good deal more), the way we kept scribbling in our journals or took our pictures; perhaps it was my Chicago White Sox baseball cap, which several of the Hanoi writers thought said SEX instead of SOX. The guy said it tickled him to think that we were paying fifty cents for the same breakfast he paid thirty cents for.

But let me ask you, where else in the world can you eat your fill for half a buck?

During the war, of course, we never ate Vietnamese food. In camp, on stand-down, we ate American-style food in the mess hall—heavy, rich, and starchy. In the field we ate canned C rations (short for Rations, Combat, Individual), which came twelve boxed meals to a case. Each C-ration meal included some kind of meat, fruit or bread, and an accessory pack containing cigarettes and matches, chewing gum and toilet paper, powdered coffee with powdered cream and sugar, salt, and a spoon. Stone-cold (grease and all) or heated with a bit of C-4 plastic explosive, it was all the same. None of the C rations looked like much, but you could always close your eyes and choke it down, and anyway, after a while, you didn't care one way or the other. But, absolutely, the one meal that made everyone gag was the ham and lima beans, which came with four crackers, processed cheese "spread," and pound cake; to this day I will not allow a lima bean in my house. I stopped bitching about the food a couple years ago when a Vietnamese veteran told me what he ate. The troops, he said, would boil up a batch of utterly tasteless sticky rice about the right consistency for sushi, and each man would take a mess of it, roll it into a sort of log-looking thing about as big around as your fist, and carry it wrapped around his waist inside his shirt. At mealtimes they would cut a slice about the thickness of a hockey puck. More than once, he told me, *more* than once that sticky rice was breakfast and lunch and dinner. More than once, he said, that log of sticky rice had to last you the month.

*   *   *

On my first couple of trips to Hanoi it became a challenge to find a cold beer. Oh, there was beer alright, Dutch-brewed Heineken and an extremely local brew in plain liter bottles, but—trust me—the only places in the world where room-temperature beer goes over is England and Ireland. The one place we found in Hanoi was built on a wharf at West Lake, which became known as the Jake McCain Restaurant; Jake short for John. Now, before John McCain was a candidate for president, before he became the senator from Arizona, he was a naval aviator. He was shot down over Hanoi on a bombing run in 1967 and spent five and a half long years as a POW at the Hanoi Hilton. He was shot down over West Lake, got busted up when he ejected, and hit the water before his chute opened. Folks from the neighborhood paddled out and dragged him ashore where (the story goes) they began beating him until a truckload of militia arrived, who I guess you could say rescued him from being beaten to death by a mob of angry neighbors.* He was so badly wounded that when he was taken to the prison hospital and put to bed, it was decided to let him die; the Vietnamese had few medical resources and John McCain was too badly wounded. Then someone did a little homework and discovered that he was the son of an admiral—his father just then becoming commander of the Pacific Fleet, no less. So, they decided that

---

* Look at it this way. If some guy who'd been blowing up "infrastructure" in your neighborhood suddenly appeared out of nowhere, what would you do? Well, on my block, we would have grabbed the guy by the very ears and kicked his ass. And when the cops showed up to bust the guy and asked what happened to him, we would say, "Well, officer, he tripped over a footlocker." That's what we used to tell the MPs after we had caught a barracks thief and stomped him before we turned him over to the battalion medics.

maybe they should fetch a doctor and see what he could do. Mc-Cain recovered, was put with the other prisoners at the Hanoi Hilton, and in 1973 was repatriated along with all the other American POWs. Years later, a monument honoring the militia who rescued him from the neighbors was erected across the road from the spot where Senator McCain was dragged ashore.

Then a restaurant happened to be built on the spot, the only place in town we found that served cold beer.

After breakfast it would be time for coffee. We walked through the neighborhood past another noodle shop, a curb-side car wash (where three guys would be swabbing down Hondas and bicycles), past a fruit stand where we'd haggle for a couple tangerines (the size of your fist) or an apple, to a little hole-in-the-wall café on Nguyen Du across from the lake. The place was run by a young couple with a small child, who shouldered his cardboard-and-vinyl book bag about the size of a boot box and started off to school right about the time we sat down. Every morning they'd set out half a dozen diminutive wicker chairs around two or three tables in the middle of the block.

Now, I'm one of those people who likes a couple cups of coffee in the morning. At the Sourpuss Café as well as most other hotels we stayed at, I would order coffee with the meal, served in a familiar cup and saucer; that was fine. But soon I'd want another cup of coffee, so the waitress would go into the kitchen and come back a long while later with, literally, *another* cup of coffee.

At any street-corner café, coffee is served a number of different ways. *Café den* is strong and hot, served in a small teacup or double-shot shot glass *in* a rice bowl of hot water with a bowl of rough-grained sugar and a demitasse spoon; *café sua* is served with sweet milk in a tall water glass; iced *café sua* is a bit chancy because you don't know where the ice has been. For a Chicago city kid, it's more than pleasant to see how social the Vietnamese are. These little cafés are everywhere, serving demitasse coffee or tea and soft drinks with fruit, desserts, or other noshy little finger food. The small wicker chairs take some getting used to, but the coffee is rich and tasty with plenty of crank to it.

Vietnam is about the size of New Mexico with a population of something like 82 million people. Hanoi is just south of the Tropic of Cancer, about the same latitude as Mexico City; Ho Chi Minh City parallels Costa Rica. It's cool enough in winter for coats and gloves and hats (there's no central heating), but in summer the heat can be brutal. So, because of the ungodly hot weather, the Vietnamese get up at 5:00 A.M. and get going, but take an extended break (or at least slow down) from 11:00 to 3:00, then work till dusk or later, *then* hang around the cafés and such.

During the dry season in summer, no one in their right mind goes out during the heat of the day unless they have an absolutely compelling reason—it's just too hot, otherwise. As an American soldier, when you first arrived in-country, the heat and humidity were merciless and it took you several months to get accustomed to it, but even so the heat was brutal and we were always pouring sweat. Recon and the line companies would go around stripped to the waist, wearing only our flak jackets, and simply thanked our lucky stars that we weren't in the ground-pounding straight-leg infantry. Those guys carried four or five canteens, or as many as they could stand to carry, along with their crammed-full pack. We always saw them with shower towels draped around their necks, and they always seemed to have that on-the-verge-of-heat-exhaustion look about them—half thousand-meter stare and half ready to collapse. We were issued salt pills, and these we ate two, three, four at a time—washing them down with canteens of water. The heavy doses of salt helped our bodies retain water, of course, but presented us with other chronic health problems like high blood pressure and the like. But what the hell, first things first, cousin. On our tracks we carried fresh water by the five-gallon can (my track had four—jerry cans we called them), and what we didn't pour into the radiator we drank or used to shave and such.

In the morning Larry and I would see people coming back from Lenin Park, groups of young girls carrying badminton rackets—playing without a net—or boys with soccer balls. The schedule seems to work something like this. Up at 5:00, eat breakfast, go to the park (the sidewalk or street) for a little exercise, hop on your bike and pedal off to work, take a long lunch (preferably with siesta), back to work till 6:00 or so, commute home, eat dinner, work at side job till bedtime, and goodnight.

Years ago I drove a Chicago city bus to pay my way through school, so sitting watching the street traffic isn't exactly the same as killing time or staring at a wall. Let's just call it semi-professional curiosity, and, of course, you just never know who is going to walk by. Even so, it's a good idea to keep in mind that when you're looking at another culture for virtually the first time, things may not be exactly what they seem at first glance.

An example: a friend of mine once spent several weeks in the home of a Hanoi family; they lived in the old "super market" part of town—on Hang Thiec, the Street of Tin, if my memory serves (sheet tin, metalworking tools, and curbside workbenches of every description); the noisiest street in Hanoi, some say. At five o'clock in the morning everyone got up—all of a sudden there was a rolling chorus of hacking and coughing, was how he put it. The radios went on. After a while he heard some very good rock 'n' roll. It certainly wasn't of the ordinary Hong Kong variety, but a very catchy, snappy little tune. Everyone within earshot turned up the volume. My friend thought: This must be the most popular song in all of Hanoi; this is *great*, a genuine slice of popular Vietnamese culture. A couple days later he learned to his considerable disappointment that the music was the theme song for the national daily numbers lottery; when the jingle came on, everyone knew that the day's numbers were not far behind.

To first look at the traffic you might get the idea that there *are* no traffic laws in Vietnam, beyond a vague suggestion to keep to the right. There is a kind of sloppy democracy in the streets that has more to do with the fact that practically everyone is on scooters and bicycles or afoot. As Vietnam has modernized and the bicycles and Honda scooters are replaced by automobiles, *that* has changed for the worse (as it changed in Bangkok, and turned that city into a horror of bumper-to-bumper traffic at all hours, endless racket, and an eye-popping pall of diesel exhaust).

For the uninitiated pedestrian visiting for the first time,

the simple act of getting across the street can be nerve-racking, but, as with all things, there is a trick. Walk to the curb, wait for a pause, a space, and simply start walking. There is no need to fret or jostle or hold up your hand, just keep moving; the traffic *will* go around you. The only city where this does not work is Ho Chi Minh, where the traffic is sometimes simply too dense to even step off the curb unless you are very nimble—or elderly (no one is going to run over somebody's mother).

In Hanoi of the early 1990s, as in most other Vietnamese cities, there were no stop signs. Traffic lights were very rare except at the intersections of the busiest boulevards of Ho Chi Minh—a thing unheard-of for generations even in the smallest American town. This chaos is typical of all Asian cities, except for the largest and most vigorously westernized. Traffic lights were so new in Vietnam that in Da Nang, near the old French colonial city hall, a light was installed along with large billboards explaining with cartoons and brief text just what they were and how to "use" them, though everyone on the street the day we were there pretty much ignored them.

The first taste I had of Asian street chaos was in Beijing (of all places), and I thought it was amazing that more people weren't killed. For instance, two six-lane boulevards would come together at an intersection where a cop stood on a circular platform in the middle. His job seemed *not* to direct traffic, but to mediate the odd dispute and to supervise and document the rare accident. These twelve lanes of traffic would merge perpendicularly—trucks of every description, buses hauling a swinging load (regardless of the time of day), vans, Beijing pedicabs, bicycles coasting along in their own broadly generous traffic lane, and (God help them) pedestrians—everything shoulder to shoulder. All that humanity came together at the intersection (like the traffic rotaries of Boston only more so), and, without stopping, just sort of merged into one large and swirling, contrary lump. No one seemed in a panic; no one was in much of a hurry. The operating philosophy seemed to be that if we all take our time and cooperate here for a minute, we'll all get through this in one piece. It was not uncommon to see a bicyclist weasel his way among the buses and trucks and for those drivers to give way without a second thought; not a thing any of us is likely to see in New York or Chicago or Los Angeles or any other American city anytime soon; pull something like that on Michigan Boulevard in Chicago at your own risk; onlookers will simply assume you are crazy. Pedestrians jaywalked through these madhouse intersections with no problem at all.

As I've said elsewhere, traffic in Hanoi, and every other city we visited except Ho Chi Minh City, seemed to move just fast enough to keep the slack out of a bicycle chain. Easy does it seems to be the unwritten law. Nobody made any sudden moves, the trucks and buses just sort of rolled along; there were no hot dogs. It's too hot to move fast, and as a general thing the Vietnamese don't look to be in a big hurry about anything. Anyway, the traffic proceeds at a nice jogger's pace—certainly not any faster than ten miles an hour. (I read or heard somewhere that in the United States the average speed of city traffic is eleven miles an hour, so Hanoi traffic is not that far behind.)

And the slower traffic does not necessarily keep to the right. There will be folks pedaling to work, prosperous shopkeepers on their Hondas poking along, fathers with children in seats bolted to the handlebars (wearing toy-store crash helmets that remind you of Tupperware bowls), and cyclo drivers grinding out the miles hauling everything from fruit-market women with their baskets brimming with produce to tourists looking up at everyone and all but grinning at the strangeness of it all. Riding in a cyclo has almost the sensation of floating.

Cyclo drivers have to be a breed apart, and I don't know that any part of their job is easy, except the wait between fares. The cyclo is the pickup truck of Vietnam and is used to haul passengers and, basically, anything that can be made to fit. Until the arrival of the French, the Vietnamese had hand-powered rickshas—that is, the person would sit down, and the ricksha guy would stand between two struts, grab a hold, and, keeping them about chest high, take off; you know, hand-powered. Then a third wheel, a bike sprocket and chain, and seat were added behind the passenger seat. Altogether it resembles a backward tricycle. By the 1930s the cyclo took its modern form, and there have been many refinements since, but few improvements. (There are some things in this world that are "design perfect"—the claw hammer, the harmonica, the book, the handkerchief, the park bench, the camera obscura, chopsticks, the lead pencil, the zipper, the paper clip, and suchlike come easily to mind.)

Across Nguyen Du Street from The Boss Hotel there were always a number of cyclo guys waiting for fares. They'd pedal up to the curb, pull the back wheel onto the sidewalk (leaving the two front wheels in the gutter), and then just sit. There seemed to be informal gathering places for cyclos all over the city. At the train station, near the hotels, around the markets, places of business, and obvious tourist attractions. I don't ever recall seeing a cyclo driver cruising for fares; if I were a cyclo driver you would not catch me cruising—the work is simply too hard for that. To look at the work, you might get the impression that it's a young man's job, but that isn't the case at all. They range in age from their early twenties to the late forties and fifties. Cyclo drivers look about as tight and tough as big-city bicycle messengers and cross-country runners; the leanest, feistiest, wiriest-looking guys you're ever liable to see, and they possess an endurance (no matter their age) that anyone who has ever worked with his back has to admire. You readily get the impression that it was men like these who strapped a couple hundred-pound sacks of rice on a bike and *walked* it those famous five hundred miles down the Ho Chi Minh Trail, unloaded, then turned around and walked their bikes back for another load—the very dictionary definition in the flesh of General Giap's assertion that the strength of Vietnam is its people (meaning, of course, the backs and legs as well as the viscera).

Driving a cyclo demands the good-natured patience of a farmer, the legs of a Hercules, an intimate and detailed knowledge of city geography (shortcuts and such), the conversational rudiments of half a dozen languages, and the negotiating skills of a big-league player's agent. You hop in a cyclo, tell the guy where you're bound, and sit back. You hand him a smoke to be friendly and get the conversation going, and, just like hopping in a cab at the airport in this country, you can get involved in an instant conversation with no trouble at all. The big downtown hotels, like the Metropole, have their own stable of cyclos—brightly painted with the hotel name stenciled prominently on the sides—but these are for the tourists come to savor the tang of *l'Indochine exotique* without having to step off the curb, and the cyclo drivers charge accordingly (referred to as the "American price"). There are no such things as cab lines, and the choice of "who's next" seems to be a matter of personal choice on the order of an emergency impulse purchase. A new cyclo with seat cushion, mirror, and other bells and whistles can cost five hundred dollars and more. There are thousands.

Of all the people I saw working and not working, I have to say that I never saw anyone, Vietnamese or otherwise, take a break quicker and loaf with more casual intent than the cyclo drivers. Between fares or during the heat of the day they would pull the back wheel over the curb into the deepest shade they could find, curl up on the wooden seat, pull their baseball hats (with corporate logos) down over their eyes, and instantly fall asleep, or pull a book out of their pockets, sit down on the park bench in front of the Light of Buddha Lake, crack it open, and read.

The first couple days in Hanoi we got squared away with the Writers Association, our hosts who had arranged the trip.

There's always a formal meeting with the big potatoes at the beginning of a trip so they can lay out what they have arranged— "schedule of work" it's called. And since there are many, many of these sorts of meetings, it's instructive to see how they're run.

You are met at the door on the street and ushered upstairs to the conference room, perhaps the only room in the building with an air conditioner in the window (most times not, though); often there are just large Hong Kong floor-fans— perhaps the only fans in the building. Down the middle of the room are two rows of low, straight-backed uncomfortable chairs, each with its own equally low coffee table. Everyone takes a seat, the Americans on one side, the Vietnamese on the other, with the head of the delegation and the Vietnamese big potato at one end (translators sit on functional stools just between). The little potatoes either sit at the other end of the room or crowd along the walls. The office women, each dressed fit to kill in their Sunday-best *ao dais,* bring in the tea things, glasses, Coca-Cola, and beer (regardless of the hour—morning, noon, or night), a bucket of chunk ice (with tongs), and plates of market-fresh fruit—usually bunches of litchi or very small Vietnamese bananas. Also on each coffee table are an ashtray and a pack of cigarettes. You can always tell how important the big potato is by the brand of cigarettes and the elegance and uniformity of the office women's *ao dais* (the most elegant were at the meetings at the foreign ministry in Hanoi and at Independence Palace in Ho Chi Minh City). The Vietnamese big potato gives a lengthy and ostentatious, boilerplated welcoming speech, adds something about the desire for good relations between the Vietnamese and Americans, and introduces the various little potatoes around the room. Meanwhile, the women make sure you drink your cup of tea, open the pack of smokes in front of you (whether you smoke or not and whether or not you brought your own), quietly and unobtrusively demonstrate how to peel a litchi fruit (there is definitely a trick to it), and fill each glass with ice and pop the top on a beer and a Coke.

When the big potato has had his say and this has been duly translated, the American head-of-delegation has his little say—hi and howdy; glad to be here; compliments and expressions of thanks for the hospitality; hopes for cooperation and solidarity between the American and Vietnamese peoples; grateful thanks for the time and trouble the big potato has gone to in arranging this *working* meeting; and introductions of the other members of the delegation. There is another round of tea-sipping and fruit-peeling. Cigarettes are offered around.

When all this is duly transacted, the actual business of the meeting gets underway. The talking, questions and answers, and translating takes forever. Sometimes the translating is not clear or the question is misunderstood, and everything has to be creatively repeated. Papers are fetched. Notes are taken. Tea is poured. Beer glasses are refilled by the women dressed in impeccable and elegant *ao dais*. Bananas are peeled and eaten in two bites. Cigarettes are lit. The meeting can go on for hours depending on the subject at hand. If this is a courtesy call, the discussion might get personal, the meeting short; in most of these where you are meeting the big potato for the first time, everyone will want to know what you did during the war. I always trotted out copies of my books and gave them as presents. It often happens that the big potato will have gifts for you (at one such meeting in the fall of 1993, Senator McCain was presented with the personal effects he'd left behind when he was repatriated twenty years before; photographs of his capture and his pilot's helmet, for instance). The meeting ends with an exchange of business cards (if these were not exchanged during the introductions—Japanese-fashion) and the meeting comes to an end with handshakes and a stroll to the door and the waiting car (or in our case, bicycles).

When the meeting is winding down, there is time for more personal conversation. There were always three questions: Where are you from, How old are you, and Are you married? They may seem highly personal to us, but the Vietnamese regard them as casual conversation. By asking if you are married, the Vietnamese are not trying to fix you up, they simply want to know about your family. Family relations are important to the Vietnamese, as in most if not all Asian countries. All sorts of social relationships are seen and spoken of in familial terms—it is not for nothing that Ho Chi Minh was called Bac Ho, "uncle," an appellation of deepest respect.*

Our upcoming tour of Vietnam by train had been known to the Writers Association members for a long while; there were many letters and extremely-long-distance phone calls back and forth. But now that we are here, specific arrangements have to be made. Contacts, phone calls, hotels, train tickets—just like everywhere else. The structure of the trip is gone over so that everyone pretty much has an idea where we will be and when. As our host, the Writers Association is responsible for us, and if there is a screw-up, they will never hear the end of it. Besides, they genuinely want us to have a good time, just as any host does no matter where you travel.

On my first trip back, in 1990, we eight writers met with Gen. Vo Nguyen Giap, one of the great, ruthless military minds of the twentieth century, and our visit with him was nothing short of amazing; the first man of flag rank I had ever met or been in the same room with. Try to imagine a group of Vietnamese writers come to the United States, asking to meet with the chairman of

* *An individual's place in traditional Vietnamese family hierarchy spills over into relationships of all sorts, and the Vietnamese sense of address reflects the social status of speaker to listener. There is no word in Vietnamese that precisely corresponds to the Western pronoun "I". A Vietnamese would speak of himself or herself in the second person as "your brother (or sister)," "your father (or mother)," "your teacher," and so on.*

the Joint Chiefs at the Pentagon. Then imagine the guy meeting them in a lavishly appointed reception hall, being abundantly cordial, showering them with welcoming gifts, a buffet, and answering any off-the-wall question that came into their heads.

Vo Nguyen Giap is an interesting cat, to say the least. Born in 1912 to a small-town scholar who had once worked in the government, Giap no doubt learned his anticolonial radicalism early when his father and friends sat around talking politics of an evening. He graduated from Hanoi University law school and made his bread teaching history. In 1941, Giap became the military leader of the Viet Minh and was thereafter a lifelong compatriot of Ho Chi Minh. Giap fought the French, the Japanese, the French again, and then the Americans; it was obvious that he knew his Vietnamese history and was a superbly determined master of guerrilla warfare. He is most famous for the defeat of the French at Dien Bien Phu in 1954, which the French did not think he could pull off, and which convinced the French to leave Indochina once and for all. He had little sympathy for the French, who had killed his young wife and child in prison and guillotined his sister in the early 1940s. When the Americans arrived, he understood that his strategy would have to include a military as well as political and diplomatic effort, and he worked from the conviction that the war might take a score of years. He understood quite well that the Vietnamese would win a war of attrition, even if it took a generation or more. He was perfectly willing, for instance, to let General Westmoreland think that the North Vietnamese attack at Khe Sanh was a big deal when it most certainly was not; in 1968, as in 1971 and 1973 and 1975, there were bigger fish to fry. He all but retired after reunification, having been a soldier at war for more than thirty years.

What on earth did we writers have in common that we could talk about with Giap? Aside from the sheer celebrity of the man (I can truly tell my grandchildren I once shook hands with General Vo Nguyen Giap), there seemed little point.

The Foreign Ministry Building—the old French colonial governor-general's residence—was kitty-corner from the Hanoi post office and within walking distance of the One-Pillar Pagoda (Chua Mot Cot, once called the Temple of Love), and surrounded by a high wrought iron fence with a cut-glass and cast-iron portico over the wide and ornate front doors. (In 1973, when Secretary of State Henry Kissinger arrived in town to negotiate the final articles of the Paris Peace Accords, it was here he stayed; at that time it had the best rooms in town.)

We arrived in high spirits, but nervous. Vo Nguyen Giap *was*, after all, a general; we *were*, after all (and everything else being equal), straight-vanilla, garden-variety grunts. We loitered in the foyer, topped with a huge glittering chandelier, waiting for the man. Now that I think about it, it was an odd juxtaposition of formality—we Americans standing inside the door waiting for him—but then, I am also certain that he was a man who enjoyed making an entrance. And when he arrived, we subjected him to a hodgepodge round-robin gauntlet of reach-across introductions. Then we were all ushered through the doorway, and through a set of sheer silk drapes that billowed and rustled as we passed, into the reception hall; two long rows of uncomfortable chairs facing each other with glass-topped coffee tables between. At the head of the room were two equally uncomfortable chairs for the general and Kevin Bowen of the Joiner Center (the "head" of our writers' "delegation"), and a smaller chair just back for the general's skinny little translator. Behind the translator on a side table buffet was a large vase of white lotuses and bright red gladiolus—bought fresh at the market that morning no doubt. And above that on the wall hung an enormous red and black and gold lacquered triptych of a farm village on a river. To the side of the room was an air conditioner the size of a Holiday Inn ice machine. On the table in front of each chair was a small vase of red and white carnations, a water glass, and a large plate of fruit—litchi and diminutive Vietnamese bananas. Giap talked about Ho Chi Minh; talked about Bac Ho's world travels. After General Giap had talked for a while, he invited us to eat something, waving his hand to all of us and indicating the fruit—virtually insisting—and picked up a short, thick banana, peeled it delicately, and ate while we continued to talk; smiling at his own gregariousness and spreading wide his arms. I sat there, drinking my tea and eating my bananas, absolutely determined to keep my mouth shut and watch. Finally he looked around the room to the several of us who'd been sitting quietly and asked (insisting in that way of all generals) if *we* had anything to ask. Well, I thought to myself, here goes nothing. I remarked that his face, his whole manner seemed peaceful (here was a man obviously enjoying his retirement, his music and poetry, his grandchildren, and his celebrity among Americans come to gawk); and why was that? He replied, "My whole strategy was peace. That is why my face is peaceful." Which is an astonishing thing for him to say, but, after all, I suppose it's another way of saying, "We won the war, bub, and liberated our country, why shouldn't I feel good about that!" There are those who will always say that the American forces (basically) never lost a battle, and that is true enough, but Vo Nguyen Giap is one of those guys who will always reply, "So what?" I asked General Giap what he wanted to be when he was younger. "When I was a boy," he said, "we 'fought' the plantation owner's children"—games the like of cowboys and Indians, I expect—"so I was *always* preparing myself for the soldier's life." Spoken like a true general, sir. All through the conversation he spoke with great humor and energy, and, it was apparent, great patience. He punctuated his talk with large swings of the arms and smiled, spoke in overbroad terms, and later, when I thought about our meeting with him, he seemed bored but at the same time performing. It's his job— meeting and greeting Americans (from big-potato journalists like Neil Sheehan to ordinary ex-soldiers like us). And if he seemed a touch arrogant (at times), let us remember that he *was* a general, accustomed to a flag officer's perks. After the meeting, after we gathered on the front steps for pictures, after our brief good-byes (the general went inside, most likely to polish off some business), I moseyed over to the general's car where the driver, dressed in white shirt and dark trousers, was napping with considerable concentration. I asked the guy if this was a good job, driving for the general. He laughed out loud and said: Yes, of course, it was a great job (how could it not be?). In any man's army, twas ever thus.

The day after our meeting with General Giap, we returned to the Foreign Ministry for a visit with Nguyen Co Thach, who was then the foreign minister; the biggest potato in the building. The morning line on Co Thach was that he was a highly regarded raconteur and would no doubt talk on and on; spoke superb English (as well as French); and would begin the conversation in Vietnamese with a ministry interpreter at his side, but had the trick of dismissing the guy if the translation wasn't going fast enough to suit him—which apparently was often. A nice piece of stage business, ask me. We were ushered into the same reception hall as the day before, arrayed ourselves among the same uncomfortable chairs, and waited for Co Thach. He arrived shortly after, blowing through the silken doorway drapes with a great flourish of arms, dressed in shirtsleeves and an open collar—coming right into our midst with long strides and a big smile, looking for all the world like a man who couldn't *wait* to meet us. (Can you imagine an appointment with *our* secretary of state, him bursting into the State Department's reception hall in a golfing shirt, loose-fitting Levi's, and snow-white Nikes, sitting down in the first chair he came to and saying, Well, hi and howdy! Let's set down a spell and just tawk! Earl, serve these gentlemen some Starbucks and some of those fresh Krispy Kremes. Here, have a banana!)

Absolutely the first thing Co Thach said was, "No protocol. No protocol" and he waved his arms like an NFL referee signaling "penalty declined." Then, ignoring the chairs under the lotus and gladiola bouquets at the head of the room, he plunked himself down in a chair among us, sitting forward on the edge of his seat—always the gesture of someone with urgent and interesting news.

Here was a man absolutely right for the job. Co Thach, very tall for a Vietnamese, had the easy manner and twinkling eyes of a porch swing philosopher. Among the stories he had was the one about his troubles in New York when he was head of the Vietnamese delegation to the United Nations. The United States would only give him a visa that restricted him to twenty-five miles. "Why is that? I was invited to speak in Minnesota and Texas," and the way he said it, a body could easily imagine him speaking to the entire state of Minnesota, or every man, woman, and child in Texas. "But your government would not allow it." As if we knew the reason and he expected us to tell him. But after twenty minutes the answer was all too clear. Here was a guy with more than enough wit and humor, intelligence and poker-savvy, and a twinkle in his eye (qualities Americans can't help but admire) to speak plainly to ordinary working-stiff Americans. Co Thach would have gotten wound up, calmly and clearly putting forth the progressive Vietnamese point of view, and—by God, great balls of fire!—he just might have changed people's minds about our image of and empathy for a Vietnamese nation united top to bottom.

Someone asked him about the economy. The subtext of it was, "Whatever happened to central planning?" Until 1986 the heavily planned economy of production quotas, price supports, state farm organizations, and such produced an economy that was going nowhere. Co Thach said that it was finally decided that "the people" could better regulate the marketplace rather than one hundred old men. This is the polite way of saying that the government leadership (who had perhaps spent too much time in the woods) was at last willing and able to read the writing on the wall, what with the Soviet Union and other socialist governments either in the tank or heading that way with all speed.

And when he said marketplace, I got the clear impression he was talking about the likes of the "super market" commercial neighborhood of old Hanoi—where the streets are named after the products they sell and the services they provide. About the Chinese, he said, "They are not teachers, and we are poor students" in the same tone of voice that your one-armed, toothless grandfather might tell you to never pick a fight in a strange bar or draw to an inside straight.

So now, Larry and I are sitting with the big potatoes at the Writers Association, talking about our plans for the trip and saying that, among other things, we wanted to buy bikes for city travel. The Vietnamese were somewhat confounded and wanted to know why. We said that traveling in an air-conditioned car with tinted windows isolated us in a way that didn't suit our needs at all; that we wanted to immerse ourselves in Vietnamese culture as deeply as our time would allow, and that bikes would give us a chance to rub up against the ordinary, working-stiff, street-corner hoi polloi (me making large gestures with my shoulders), a level of contact simply impossible from a car.

Our hosts at the Writers Association were both pleased and horrified by this. As we moved around the country, more than one person intimated to us that traveling on bikes was undignified for foreigners, and I suspect that Tu Nam, head of the Writers Association, and the other big potatoes felt this way, too. But they also said that street traffic was *dangerous* and repeatedly admonished us to please, please, *please* be careful; basically because if anything happened to us, they would have felt more than rhetorically responsible.

And when they got used to the idea of Larry and me on bicycles, we told them we wanted *Vietnamese* bikes. "Good Lord," about sums up the look on their faces.

Actually, the bikes were an adventure all by themselves.

After the meeting, we went shopping at a state store across the square from the old French opera house advertising a Haydn-Beethoven concert with an enormous red banner. We wanted *Vietnamese* bicycles. The Vietnamese themselves prefer Chinese-made bikes with brand names like Flying Pigeon and Lion—millions of bikes virtually indistinguishable; like Model T Fords, you can have any color you want as long as it's black. The bikes are smuggled across the Chinese border at Lang Son, though "smuggle" is a too elegant word; all you had to do was walk across the border, buy one, and ride it back. I don't know that you could call *that* smuggling.

The place was one of those artifacts left over from the French Colonial days and had the look of an old Sears store, or maybe the Woolworth's "mother store" in Cheyenne, Wyoming; large, square central pillars; ironwork mezzanine rails with other departments above stairs; glass-topped display cases; old-fashioned pigeonholes for boxes of stock; dim lights—there was actually more light coming in from the street. It is easy to imagine extremely bored French provincial housewives in '30s and '40s dresses and big hats and sweating like pigs buying a dollar's worth of ribbon and having it delivered to the house. The bikes were packed together in the middle of the space, showroom fashion. There were numbers of models in different sizes and a range of prices, but nothing with many "bells and whistles."

We should have known better. We must have been out of our minds. These things were always breaking down.

Yeah, we should have known something was up by all the dust on the seats, and the fact that most of the tires on the two hundred or so bikes were flat. But we absolutely insisted on buying Vietnamese bikes; it would support the "economy," give a couple guys some work, on the street we'd be just "regular folks." We tried a couple out, everything seemed to be loose. To make a long story short, we bought fifty-dollar bikes—definitely the American price, and way beyond what anyone else would put up with. Kim Hoa took our greenbacks and went to change them into Vietnamese dong, and came back a while later with a purse full of money.

Kim Hoa whipped out an entire handful of dong, very deftly and quickly counted out these minibundles—her fingers just flying—she'd whip through 50,000 dong (in 5,000 VND notes) in a flash—and set the minibundles on the counter. The sales-woman counted them again (everyone keeping an eagle eye on the cash), rewrapped them with the rubber band, put them in the cash box, and then carefully wrote out a receipt.

While money was changing hands, the shop guy tightened all the nuts and bolts and pumped up all the tires. We had them throw in a couple locks—a handy item, even in a socialist coun-try (better safe than sorry); as a general thing locks only keep an honest man honest. By the time we rode back to the guest house, a matter of half a mile or so, not only was Larry's tire flat, but when we took it to the local street corner bicycle me-chanic across the way from the guest house, he allowed as how the whole gizmo, tube, tire, and all, was a total loss. The guy fixed it but it didn't hold, so the next day Larry loaded himself and his bike into a cyclo and off we went to the store for some well-considered satisfaction. After some hemming and haw-ing, a new bike was produced, the nuts and bolts tightened, the seat adjusted, the tires pumped, and off we did pedal; and this model lasted Larry the whole trip.

Take it all around, riding the bikes was a great kick, part transportation and conversation piece. We would be cruising along and folks would simply get up next to us. Americans on bicycles? We're writers; Larry's a poet; we're here to see the country. Soldiers? Well, yes; you? Well, yes; I was *bo doi* in Quang Tri. We would be invited to his house for tea, for beer, to meet his family and simply talk; the Vietnamese were as eager to talk to us as we were to talk to them; there were many exchanges like that. Guys who had fought at Dien Bien Phu in 1954; guys who had been at China Beach in 1965 and watched the Marines come ashore to be met by delegations of young Vietnamese women in *ao dais* there to welcome them with flowers; guys who fought the 1st Air Cav in 1965 at the Ia Drang Valley; guys who fought again and again in the A Shau Valley near Khe Sanh; guys who fought right up to the last day in 1975, like Bao Ninh; guys who fought at places we never heard of, but had the scars. And when the bikes broke down, we pulled up to the next curbside bike mechanic, pointed to the problem, and the guy would go to work; a crowd would soon gather, the kibitzing commence, cigarettes passed around, and, among other things, we would see how things are done the long way. Fifteen minutes later the bolts were tight, the tires pumped, money paid, a tip of the hat and a wave, and we were on our way. There were, you may be sure, many of these encounters. Next time I do this, I will definitely get the Vietnamese-made Japanese bike.

One night Larry and I took off on our bikes for an evening of water puppets. There was a theater company out in the south neighborhoods. The trip took ten minutes and took us into a part of town where, it was clear from the looks and stares (and downright gawking), foreigners don't often go. At the theater we paid a nickel to park our bikes. You pay your nickel, the guy gives you a numbered chit and writes the number on the seat with a bit of chalk. After the performance, we fetched our bikes and rode back to the guesthouse through dark streets (minding the potholes). I mention this because for all the times I traveled on foot or aboard a bicycle at night (in Hanoi at least), I never felt that I was in danger of being bothered. Not that there were cops at every corner; in fact, we saw few on the street. It's just that the atmosphere of the place simply didn't convey dread in the way that the streets of an American big city does.

Water puppetry of northern Vietnam (more properly Tonkin) is a unique traditional folk art form of the "village pond"—a euphemism for the common people, presumably working-stiff farmers and other small-town folks. The first recorded water puppet show was performed in 1121 at the celebration of the emperor's birthday; which means, of course, that the form goes back much farther than that.

It is a theater form that could only have come from a wet-rice culture. The performances are traditionally done on the surface of a rice paddy or shallow pond, along the shoal water of a river or creek; outdoors and at dusk, so that the natural surroundings of the village, the deepening colors skyward, and the weather drifting along contribute unexpected dramatic effects. A hut is mounted on stilts and faced with a bamboo scrim behind which the puppeteers stand, and in back of that is a shed to store the puppets. The theater in Long Tri Lake at the eleventh-century Thay Pagoda is the only extant antique puppet theater left.

Traditionally, puppeteers were the village men, and the skills and stories were passed down from father to son. The puppets are carved from the ankles up of fig or other light wood, painted with lacquer, and sometimes covered with touches of gold or silver foil; doll-sized but often quite hefty

looking. A puppet is attached to a long bamboo pole and manipulated by means of a long piece of string or fishing line. The
puppeteer stands behind a scrim knee-deep in water (about
the depth of a rice paddy), and, slightly stooping over, holds the
bamboo pole out of sight and works the fishing line; heads bob,
arms bang drums or wield swords, raise and lower baskets
catching fish; animals undulate their bodies as if swimming.

The traditional story repertoire includes brief scenes of
village life common to the audience: farmers and fishermen,
woodcutters and boatmen, weavers and wrestlers, peasants
whitening rice or pounding it to flour; white-haired, scaly-
skinned geezers and witchy hags, wild children and cruel ban-
dits, and clodhopper young bucks and beautiful daughters
(mindful of family duty). Other stories involve recreations and
games, fables and legends involving animals—buffalo, turtles
and hares, and phoenixes; cock and buffalo fighting. Still others retell stories and scenes taken from popular and classic literature, or significant historical events; the Trung sisters'
unsuccessful revolt against the Chinese in the first century,
Emperor Tran Hung Dao's defeat of the Mongols three times in
the thirteenth century, Le Loi's defeating the Chinese and es-
tablishing the Le dynasty, which lasts until late in the eighteenth century, and others.

The performance begins with a peal of firecrackers min-
gled with the call of a horn. The crowd gathers along the bank
and settles down. The performance itself is accompanied by
music and more fireworks—the water bubbles and stirs with
gunpowder smoke, roiled by the swishing of the puppeteers'
bamboo poles, dunking and splashing as puppets appear out of
nowhere and just as quickly disappear as if swallowed up. A

performance ends with more noise and a loud finale of music—
like fireworks night at Comiskey Park.

Before the end of World War Two, water puppetry was left to
the peasants, a folk art if there ever was one. Now it is regarded
as a bona fide theatrical art form in its own right. There are
several professional troupes in Hanoi, though these performances are done indoors in what looks like a knee-deep wad-
ing pool.

In all weathers and in all seasons, an evening of water puppet theater is a thing to behold.

As I've said before, Larry and I bumped into many a surprise,
but thinking back on all my trips to Vietnam, perhaps the most
strange was meeting novelist and former secretary of the Navy
James Webb, one of those professional patriots I thought I was
safe from—meeting, I mean. I never thought I'd run across him
in a million years, but absolutely and irremediably the last
place I thought I'd ever bump into him was Hanoi. I cannot
think of anyone whose point of view about the war in Vietnam
could be more opposite mine; I don't know that he and I could
agree on so neutral a topic as the weather.

We ran into him at an evening of traditional Vietnamese
theater we were invited to attend, compliments of the Writers
Association. The theater was right next to the Writers Association offices, down the street and around the corner from our
guest house. We rode our bicycles, though it was close enough
to walk. The streets were plenty dim, and making our way with
the traffic (mostly bikes and a few Honda cycles) was accom-
plished almost by touch.

To explain the full extent of the evening's particular bizarreness, let me back up a moment.

Larry and I flew into Hanoi on a plane much less than half full; Vietnamese government people (you could tell by the blandly straight-cut, high stiff-collared, iron gray tunics), Japanese and Thai and Hong Kong businessmen, a smattering of tourists like us, and the odd German and Scandinavian on a lark. As we boarded the plane, one guy—about our age, trim looking with an Izod pullover shirt and a zip-a-dee-doo-dah Boy Scout haircut—caught my eye. I nodded "hi and howdy" to the guy and sat down, and didn't think another thing about him. Still and all, he looked familiar, but then so do a lot of people; how many guys who work in the pit mines of Pennsylvania and West Virginia do you suppose look like Charles Bronson or Mike Ditka?

Well, a couple nights later at the theater, who should sit down across the aisle but this cat from the plane. I glanced over and said, "Fancy meeting you here," meaning that Americans of a certain age are rare in this neck of the woods, and that bumping into someone a second time (even though we'd only exchanged a glance across the airplane cabin) in a large city is rarer still. He looked at me with a peculiar expression on his face, fetched himself a long breath (as if trying to think of something to say), and finally blurted out, "Do I *know* you?"

He said it like someone would say, "I thought you were in *prison*" or "My wife heard that you were *dead!*" or "*You're* the son of a bitch who totaled my Caddy with your bus on the Outer Drive in '68." (Yeah, bub, that was me, alright.) I introduced myself and extended my hand across the aisle; whatever else

my mother taught me, she taught me to be polite. We shook and
he said his name was "Jim Webb."

It took a solid minute for that to sink in, but then it dawned
on me—oh! *James Webb*; wrote *Fields of Fire* with the author's
photo of that hotshit Marine Corps 1st looie sinking his teeth
into a cee-gar with a well-weathered, well-crushed bush hat
down over his eyes John-Wayne, cavalry-to-the-rescue fashion (his novel written for people who think that the war was a
sad undertaking to be sure, but really a *good* thing, for all that);
in other words the look of a ticket-punching lifer who couldn't
wait to get everybody killed. When the novel first came out, I
tried to read it, but the writing was so bad, the story itself so
cliché ridden, that I finally just gave up; nowadays I expect you
could say that it was like reading Tom Clancy without a net.
Toward the end of the story the main character looks on the
VVAW antiwar demonstration in Washington and simply cannot believe that the veterans are, indeed, soldiers (they must be
lying; poseurs; impostors); as if the character simply cannot
get his mind around the fact that anyone would oppose the war,
much less sharply act upon his opposition with any eloquence
or conviction; the larger, the deeper, more urgent questions
about our involvement in Vietnam seem not to have occurred to
him. In the early 1980s, when Maya Lin's brilliant design for
the Vietnam Veterans Memorial was first announced, Webb
took one look at the model and said something like, "This
means trouble"; he had apparently been expecting the usual
hero's hokum that is not interested in the connection between
a man's senses and his soul (and the accompanying spiritual
upheaval, not to say betrayal of spirit); it has always been clear
to me that James Webb's imagination could not cope with the

memorial's clean and simple audacity. Later, he worked in Ronald Reagan's Pentagon as secretary of the Navy and was a booster for a 600-ship Navy (and what the hell were we going to do with 600 ships?). In short, a lifer's lifer, one of those inside-the-beltway, Washington guys—always hanging around the fort with his hand out. To be fair to Webb, I don't suppose he had much use for me, or Larry Rottmann, either.

I wonder whether my expression was as brackish as his when we shook hands across the aisle, and what impression we made on him. I introduced him to Larry (who had been loudly and bluntly active in the VVAW, remember); it was clear to the three of us that our reputations had preceded us into the room.

We asked him what he was doing there. He said that he was working for the Disabled American Veterans, supplying the left-behind wounded ARVNs with prosthetics (a project that produced actual results, diplomatically and otherwise). He asked us what we were doing there, and we told him that we were there to horse around, ride the trains, and take some pictures. I think he scoffed or snorted or made some other thoroughly irritated gesture—a sort of shivering all over as if his bowels were trying to move a watermelon. Anyway, he turned eyes-front and spent the rest of the evening staring straight ahead; I don't know that I've ever seen anyone enjoy an evening of theater less than he.

The evening consisted of a series of one-act skits performed by a company of six or eight actors and four or five musicians who sat on the floor to one side playing traditional instruments. One musician readily and enthusiastically kibitzed with the audience—like you'd expect to see at a performance of children's theater, though there were very few kids in the room; he was very animated and demonstrative, and I suppose his gig was to cue the audience to what was happening and offer verbal editorials. Children's theater is always a hoot, because kids have no problem suspending their disbelief and are just as likely to cheer the hero and boo the villain as not.

The only piece I recall was a very good version of the Cinderella story, which the Vietnamese know as "The Story of Tam and Cam." The prince in gaudy, traditional mandarin dress; the stepmother and sister in circusy *ao dais*; Cinderella, of course, in pathetic and ashy peasant rags. After the palace ball, the prince and his flunky (famous slipper in hand) come to the house of the nasty stepmother; the slipper is presented; the stepmother takes her daughter into the next room with the slipper; the foot is, of course, *too* big; mother produces a dagger about the size of a Scottish Highlander's claymore sword from the sleeve of her *ao dai*; says, "Cut off thy toe, when thou art Queen thou wilt have no need to go afoot"; the girl produces her foot and sticks out the big toe; *whack!*; off comes the toe with one stroke; she swallows the pain, *squeezes* what's left of her foot into the shoe, hobbles out to the prince, and off they go; they pass under the tree growing up from the grave of Cinderella's mother; two birds sing a little ditty (done by the musicians stage-right) that tells the prince that this is not the true bride; oops; back to the house they go.

Cinderella comes forward and asks most humbly to try; the other women scoff and sputter and protest (the drummer banging loudly and shaking his head extravagantly at the audience with a very sour expression; Evil stepmothers are such a pain!); Cinderella produces her foot; the slipper fits like a glove. Voila! The true bride found! Hooray! Everyone but the stepmother and daughter lives happily ever after. In another version I heard years later, the stepsister one day asks Cinderella how does she maintain such beauty, and Cinderella tricks her into jumping into a pot of boiling water, makes a soup that the stepmother then eats with "much relish" (my Hue storyteller said), but when she finds out her daughter is in the soup she drops over dead.

Throughout the evening I couldn't help but glance out of the corner of my eye, looking Webb up and down. He had the uncomfortable, squirmy and sour, pained expression of a man who had just sat on a fork, but was too tight-assed or polite or *some*thing to reach around and pull it out. Not only did he hardly move a muscle, he looked like a kid sitting in church wearing a brand-new wool suit (collar cinched tight with a polyester four-in-hand regimental tie), sweating and itching like crazy, and knew if he moved a muscle his old man would reach around and pound him one *good*.

At the end of the evening I looked over and Webb was obviously long gone; no doubt as eager to be rid of us as we were of him.

Months later Mr. Webb published a story of his trip in *Parade* magazine, that ubiquitous Sunday supplement. It is revealing of his character that of all his travels in Vietnam, all the things he may have seen, that the one moment that sticks in his craw with singular, irritated clarity is his departure at Tan Son Nhut Airport. When he got to customs with his baggage and his paperwork, the guys took one look at him and told him in no uncertain terms—in that way of all cops—to open his suitcases for inspection. You can imagine him looking very sternly at the guy when he said, "Do you know who I am?"—certified war hero and professional ex-Marine, former Pentagon official and one tough cookie; yes, yes, James, we know—husband and father, homeowner and taxpayer. The customs guy, we can imagine, looked at him and said, "Yes, Mr. James Webb, we know exactly who you are. Now, crack open your bags and be quick about it, and let's just have us a look-see."

The Sunday that we were to leave Hanoi for Vinh, three hundred kilometers south, Larry and I got up early to mosey over to the railway station on Le Doan Street (Highway #1) and take some pictures. It was only half a dozen blocks from the guest house. The railway station was one long, tall French-built affair, with a mansard roof, that stretches for several blocks; both the station and home office. The building was obviously bombed during the war and repaired without being restored. The façade of the waiting room was hideous and not at all in keeping with the spirit of the overall design of the building. There is a big sign across the front—Ga Ha Noi (Station Hanoi). The cyclo drivers lounged around the gravel yard, waiting for fares; the fruit sellers napped behind their piles of oranges; the freelance redcaps sat around, waiting for work. We went up the steps and inside. It was dim and dusty; the only light came through the doorways and tall old windows; by now it was clear that no one turned on the lights of any interior of any building until it was absolutely necessary. We walked across the old-fashioned, high-ceilinged waiting room to the door leading to the platform, but it was locked with a thick length of chain and a brass railroad lock. We rattled the doors to get someone's attention, as if two tall Americans weren't going to draw attention enough. And sure enough, very quickly the rattle of the glass brought several uniformed railway employees out of the stationmaster's office; young women in railroad uniforms, Vietnam Railroad insignia, badges, hats, and all.

No one was allowed to go out on the platform before train time.

We indicated our cameras with utter guileless innocence, reiterated our very Americanness, and tried to explain in slow and carefully enunciated English (on the chance someone speaks English) that we were humble tourists and simply wanted to take pictures of their magnificent station.

No. No, they said. Sorry.

What was the problem?

You must have a picture-taking ticket (at least that's the gist of it), one of the young women said, pushing back her trainman's hat.

This innocent misunderstanding went on for some time. We rattled the lock, pointed to our cameras, and shrugged our shoulders, guileless tourist-fashion. Soon a man came down the stairs and crossed the waiting room to us; he spoke high school English.

Was there a problem?

We wanted to take pictures, that was all!

Oh, yes. Take pictures. Yes, fine. You must in that case buy a picture-taking ticket. It is five dollars, U.S.

Five dollars was 50,000 Vietnamese dong, a considerable sum. The guy was smiling and spry for so early on a Sunday morning. He put his hand on my shoulder and grinned big—five dollars is not so much in America, he says.

We showed him our train tickets for the 9:00 P.M. second-class train to Vinh.

Well, yes. These are *train* tickets. To take pictures you must buy a picture-taking ticket, and five dollars is not so much in America, he says, all palsy-walsy.

There was no dissuading him. We wanted to take pictures; we must take pictures; so, pay we must. The whole gag was beautiful to watch; the kid in a coat and tie, and big hair; the several women standing aback, eyeballing the whole encounter.

Absolutely no problem, we said, and were ushered into the back of the ticket office. One of the uniformed office women sat us down at a desk. She reached in a drawer for a sheaf of "tickets" and a sheet of carbon paper. Another of the uniformed office women brought us a pot of tea and two cups.

Tea?

Yes, thank you.

Where are you from? How old are you? Are you married? Soldiers? Take our picture?

The young woman behind the desk slid the carbon paper between the first two forms and began writing. "Admit one American Tourist onto platform for pictures." Signed, and date-stamped; the tissue paper dry in my hand. We took the tickets, turned our hats back, asked everyone to step out to the platform, stood everyone in front of a kiosk, told everyone to smile, and took their picture. The five bucks would buy everyone working in the station breakfast and lunch for a week.

Larry and I spent the afternoon in the old part of Hanoi, lollygagging on our bikes. We had a bit of lunch and then took a bit of a walk through the park around the lake there; Hoan Kiem Lake, or Lake of the Returned Sword. The lake is attached to the legend of Le Loi, a fifteenth-century royal revolutionary who defeated the Chinese and established the Le dynasty. The story goes that while fishing one day, he pulled up a golden sword that gave him supernatural power. He gathered an army, fought and whipped the Chinese, declared himself emperor, and established his capital again in Hanoi. One evening he was cruising Ta Vong Lake, as it was then called, for the cool, when a golden tortoise swam up and asked Le Loi to return the sword to the master of the kingdom of waters. Le Loi gave the tortoise the sword, which it swallowed whole, and swam away. And ever since then, as the story says, the lake has been known as Hoan Kiem, or Lake of the Returned Sword.

4

Going Down the Line

When it's time to get a move on, you have the choice of three transportation systems. Vietnam Airlines, the roads and highways (most notably Highway #1), and the Vietnam Railway.

If you're in a hurry—and time is *always* money—go by air, but you will come away thinking that all of Vietnam is city life, and the rest a blur of rice paddies, creeks and rivers, desolate scrub, lumpy mountains, and spongy-looking jungly woods. In other words, you won't see much at all.

The road system is another matter. Highway #1 is the Oregon Trail, the Lincoln Highway, the Route 66, and the Interstate 80 of Vietnam; the country's main drag not any wider or more grand than a decent country two-lane blacktop road. It stretches the length of the country from Lang Son on the Chinese border to Hanoi, south through Ninh Binh, Thanh Hoa, and Vinh in the north, and then to Hue, Da Nang, Nha Trang, Phan Thiet, and Ho Chi Minh City. From Ho Chi Minh City the highway turns northwest to Cu Chi, Trang Bang, and Go Dau Ha on the Cambodian border, and on to Phnom Penh (see map). The highway was begun a thousand years ago by the kings of the Le dynasty and called, of course, the Imperial Road, though it didn't amount to much more than a cart trail. The French developed it to something more or less suitable for automobiles and eventually made a stab at paving it. And, of course, when the Americans arrived, they *poured* money all over it; nothing says money and progress to an American (most often in the same breath) quite as much as a grand public works project. A free-for-all of trucks and intercity buses, cycles, and the occasional automobile move along at a round and solid 30 or 35 miles per hour, top end, negotiating through an endless gauntlet of pedestrians and bicycles, oxcarts and handlers on foot, and equally hasty-minded oncoming traffic—everyone honking their horns to a fare-thee-well.

Highway #1 is also a thousand-mile workbench, promenade, and hangout.

Farmers spread their harvest along the paved edge, turning it over with hay rakes. The warm and hard tarry surface is perfect for drying everything from rice and paddy hay to coffee and tea to broom straw and hemp. Sugarcane is stacked and cut for bundling. Bamboo is split. Baskets weaved and sorted. Logs piled and graded, peeled and milled. Coconuts husked. Straw woven into mats. Sheaves of purple firecracker paper dyed and dried, rolled and loaded, and strung together in bunches like garlic. Brick-makers stack and sort. Stonemasons bang out building stone and statuary. Mechanics tinker with machinery. Carpenters dress planks and build furniture of every description. Cattle and ducks and geese are driven to pasture. Water buffalo browse along the shoulder and amble at their leisure to the fields. It is not uncommon to see a large truck of one kind or another with an axle jacked up along the side of the road, and the tire rolled off to town for a patch while the driver takes his tea in the café across the way; no sweat, the tire will take all day. The farther south you go, the more large ex-American trucks you see, and, funny thing, on the cab roofs are mounted fifty-five-gallon drums with a tube running down into the engine compartment; apparently the water pumps were the first things to go on the trucks, and rather than replace them, you simply mounted a tank on the cab roof and ran a line down to the radiator; this explained the trickles of water on the road. Refilling the tank was not a problem.

Highway #1 is also playground, soccer practice field, and badminton court, and, because the road might well be a village's only thoroughfare, it is also town square, festival center and dance floor, morning market and bus stop, wedding aisle, funeral procession path, café, and all-purpose meeting place where folks can loaf over a pot of tea, read the newspaper, mingle and gossip, swap stories and news, watch the traffic, and nap through the heat of the day. Drive the highway far enough and you will have to contend with all this and more, and it can be (by turns) exhilarating, nerve-racking, and exasperating; there are few stretches of highway where a body can "make time"—that quintessentially American, long-haul, over-the-road compulsion of ours. Travel far enough down Highway #1 and you're liable to see anything and everything, moving at a sedate and dignified Vietnamese pace—if not standing still.

I traveled stretches of Highway #1—around Hanoi and Vinh, between Da Nang and Hue, and between Ho Chi Minh City and Tay Ninh. And I've traveled it in Honda tour vans and four-door sedans, even by motorbike, but I never had the nerve or the stamina to go much farther than that.

The French-built Vietnam Railway parallels the highway for long stretches, and for my money there is no better way to *see* the country (or any country for that matter); the train and the four hundred people aboard form a moving village. I'm one of those romance-of-the-rails guys, anyway, and as I've said before, riding any train is a great kick; that was half the purpose of our trip—and whenever I travel to Vietnam I always manage to take the train.

Beginning in the 1880s, the French took thirty years and more to build it. A good deal of the right-of-way was destroyed during the French and then the American wars, especially in the south, where it was all but obliterated. After the war, the railroad was rebuilt in about a year's time (pretty quick work, ask me), and in April 1976 the first train in twenty years made the trip from Hanoi to Ho Chi Minh City. That trip took five days, but then this was the inaugural run of the Reunification Express, so it was obliged to stop everywhere for festivals, celebrations, and parties all along the way; a very big deal. Nowadays, the first-class train makes the trip in thirty hours—not bad.

Back in the 1970s, travel writer Paul Theroux made a rail journey from London to Osaka and back.* He stopped in Vietnam, the war still grinding along, just long enough to ride the

---

* *Theroux's book* The Great Railway Bazaar *is a classic of twentieth-century rail travel as professional sport, though you get the clear notion that traveling with Mr. Theroux was not much fun.*

trains. "People have done stranger things in that country," he said of his visit. Amazingly, during the war there were stretches of rail still open. Specifically, he rode the Saigon–Bien Hoa market train; not twenty miles, perhaps the distance from Union Station here in Chicago to Highland Park on the North Shore. The Saigon stationmaster made his private car available to Mr. Theroux and coupled it to the rear of the third-class, one-size-fits-all market train; when Theroux walked through to take a look at the hard-seat coach passengers, he was struck by their seemingly endless resourceful forbearance—of which, by 1973, the Vietnamese had in abundance (though not for long). Then he traveled up to Da Nang and rode that station-master's car for the five-hour trip over the Hai Van Mountains to the old imperial city of Hue. And afterward he wrote that "of all the places the railway has taken me since London, this . . . ," the rail line rising steeply and winding into the mountains along the coast, then down to the seaside fishing village of Lang Co and north across the flatlands and into Hue, ". . . was the loveliest." And, Mr. Theroux, I couldn't agree more. It is something undeniably enchanting to roll along with the mountains rising in front of you out one window, and the broad panorama of the South China Sea out the other.

Sunday evening, Larry and I made our way to the train sta-tion for the 9:00 train south. We had to take the second-class train because, for some reason known only to the railway, our bikes were not allowed on the first-class train. Oh, well. We had to be the only foreigners aboard, and the conductor had obvi-ously been told that a couple of Americans would be traveling on his train because he greeted us warmly at the platform, had a couple of the whippersnappers on the crew take our bags, and personally escorted us to our compartment; he even shagged a couple guys out so we could have the compartment to ourselves. The compartments were functional in the extreme—certainly only a train buff could love them—and amounted to a kind of camping; small fluorescent light overhead, four bunks lined with a thick straw mat and one pillow, a swing-up table next to the open window, and a door that was always kept open for the breeze. At 10:00, precisely an hour late, the train pulled out and rolled at a walking pace through the station and onto the main line. Immediately a young railroad cop dressed in extremely casual street clothes appeared in the corridor and sat in the window across from our compartment, looking at us and then looking away (the way cops do). He had a vintage Colt .45-caliber semiautomatic pistol stuck in his front pocket, and by the pinched expression on his face, the pistol obviously made sitting on the windowsill uncomfortable for both of us. Cops and guns make me nervous; this goes back to when I was a kid hustling yard work door to door; the cops would hassle me, seemingly just for the entertainment (the way cops do).

After we got a mile or two down the line, the conductor paid us a visit to schmooze and practice his English. I asked about the cop, sitting just there in the window as we cruised along into the wide countryside—Highway #1 just over there. The conductor told us that the guy was there for our "safety." Yes, yes, I said, but why is he sitting *there*. The conductor repeated that he was there for our safety. A third time I asked, and a third time we were told that the man was a railroad cop and that he was posted just there outside our compartment for safety's sake, and we finally let it go at that. (Lesson number one, Larry: don't pester.) Only later in Ho Chi Minh City did a friend of ours explain that since there were stretches of track where the train slowed to a walking pace—the trackage was so poor, the elderly bridges so delicate, the roadbed still needing to be up- graded to something resembling Class-A—that gangs of ban- dits would swing aboard and rob everybody blind.

Oh. Well, in that case, Mr. Conductor, good plan and many thanks.

Larry and I talked with the conductor until the conversa- tion petered out; it was late, he said; he wished us a good evening and left. Larry and I sat at the open window a good long while, entranced with the countryside sliding sedately past, but after a while even we had to admit that we were tired and turned in. The kid with the gun sat in the window, bored and mindful, and glad for the breeze.

I awakened a couple of hours later, groggy and sore from sleeping on a straw mat—the best the train had to offer—but too tired for sleep. So I sat around, poured some hot water from the thermos for tea, and leaned over the wide sill as the train slow- cruised along beside Highway #1. I watched the farms and vil- lages; the little dots of light from house lamps blinking through the hedgerows and bits of woods; and broad stretches of rice paddies bathed in moonlight (not quite full) the color of old silver. And there it was, the country at peace, the thing I had come to see. Just then I had the strong sense that I could reach out and touch all this in the same way that you can stand at the edge of a high place and have the strong, undeniable urge pass through you to raise your arms and step forward; always the urge to be resisted, mind you. I don't mean to put the inference

of menace into that moment, but, I tell you, that all-at-once
texture, that all-at-once impulse, stayed with me the rest of the
trip.

During the night we passed over the famous Ham Rong
Bridge near Thanh Hoa. During the war the bridge was called
"the Dragon's Jaw" because of the many antiaircraft guns arrayed along the shores of the Suoi Ma (River of the Horse) to
protect it from the endless American attacks. Such a bridge was
called a "choke point" and was attacked with bombs, rockets,
and even "Walleye" missiles. The Vietnamese were no less determined to defend it. During the war the bridge was destroyed
and rebuilt many times, and something like seventy planes
were shot down at that place.

We arrived in Vinh at five in the morning, just as the sun
rose, and rode our bicycles to the city-owned Friendship Hotel.
The yet-to-be-installed elevators—the boxes and cables, the
chains, and other hardware—were laid out all over the lobby,
and someone had nailed two-by-fours across the open elevator
shafts to keep people from falling in.

Vinh had the look of a hardscrabble town, not unlike Da
Nang, which reminded me of Gary. During the war, Vinh was
the jumping-off place for the Ho Chi Minh Trail and was heavily bombed from the beginning of the war to the end. The trains
and trucks came south; the goods were unloaded and transferred to other trucks, which then made their way to the trail,
and south they went. It was at Vinh that tons of rice by the
hundred-pound sack were loaded on bicycles. It worked this
way: the seat is taken off the bike; a bamboo pole is tied to the

vertical strut, and another pole is tied across the handlebars; a hundred-pound bag of rice (or two) is tossed over the horizontal crosspiece and secured; then the guy takes hold of the two bamboo bars, joins the rest of his transportation company, and together a couple hundred guys walk a couple carloads of rice west to the trail and then south, dodging American air strikes all the way. The journey took months. The tail end of the trail along the Cambodian border was straight north of Tay Ninh City. The supplies would be unloaded along the way and distributed, and the bicycle guys would turn around and go back for another load.

It just so happened that we were in Vinh for Ho Chi Minh's birthday, and his childhood home just happened to be west of town. The village of Kim Lien—known locally as Lang Sen, the Village of Lotuses, Lotusville—was surrounded on all sides by lotus bogs, and that pastel aura of flowers and fragrance was everywhere; what a place to live a childhood.

Ho Chi Minh was one of the great political figures of the twentieth century and a genuine hero of Vietnamese history; this, in the same way that soldier-statesman George Washington is a hero of ours. In fact, the Vietnamese fighting for the revolution understood quite clearly the comparison between Ho Chi Minh and Le Loi, who defeated a large and unwieldy Chinese army with hit-and-run guerrilla tactics in the middle of the fifteenth century. (The Chinese were hardly seen again in Vietnam until 1945, when they, along with the British, waltzed in to secure the country from the Japanese after the war and waited for the French to return. When the Chinese left, they took everything with them that wasn't nailed down.) Ho Chi Minh was born Nguyen Sinh Cung in 1890, and over his lifetime he used many, many pen names and noms de guerre. I have a sneaking hunch that the name gag was a kind of game with him, though he always picked a good one; "Nguyen the Patriot" as a young radical in Paris just after the Great War (educated and passionate, but naïve), "Ho the Enlightened" (still passionate, and wised-up plenty) when he returned to Vietnam in 1941 to lead the revolution.

By 1911 he was ready to leave Vietnam to travel, study, and organize. He would be gone thirty years.

In this he is very much like the Indians of the Great Plains in the late nineteenth and early twentieth centuries who purposely exiled themselves in order to have a look at the White Man for themselves. Ogalalla Sioux holy man Black Elk, Chief Sitting Bull, and others joined Buffalo Bill Cody's Wild West Show and toured the eastern United States, England, and Europe. At a command performance for Queen Victoria and her court, she thanked them and then said that if they were *her* people, she would not treat them so—perhaps forgetting what the British had been doing for a couple hundred years in Ireland. What Black Elk and the others saw was mighty discouraging; there would be, it was clear, no end to the deluge of White Men, and resist as they might, it was also clear (as Chief Seattle once said) that the Indian's night was going to be long and dark.

But Ho Chi Minh was not discouraged by what he saw, and he knew (along with many other shrewd and intelligent, supremely patient Vietnamese) that French colonial rule was defeatable. The Vietnamese had done it before, they could certainly do it again; Ho Chi Minh used the phrase "putting your

thumb to the earth"—and how many times had they thrown the
Chinese back over the fence?

He traveled the world, learned five languages, watched and
listened and learned the way of the world. During those first
years he worked hotel kitchens in Paris, New York, and Boston
(as a pastry cook, some say), and elsewhere; imagine him scut-
tling around the bakery ovens of the Parker House in Boston,
dusted head-to-toe with confectioner's sugar and soaking up
all the hotel-kitchen gossip. After the Great War, he showed up
at the Versailles Conference to talk about Vietnamese independence—and was promptly shown the door. He was a founding
member of the French Communist Party. Photographs of him
during those years reveal a slight and boyishly handsome young
kid; what my toothless, one-armed grandfather called "skinny,
run-down, and nervous." Perhaps he didn't photograph well,
but then I think he was one of those rare men who get more
beautiful as they get older. Tolstoy was just plain plug-ugly until he'd had half a dozen kids and written a couple books. In his
later years, Sam Clemens sported a wild head of stone-white
hair that well showed the twinkle of bemusement in his eye at
the silliness of the human race. Here in Chicago we have Studs
Terkel, master raconteur, and sitting down with Studs is like
hanging out with your favorite loopy uncle because you know
you'll get the straight story, the one that the rest of the family
doesn't want you to know.

Let's just say that Ho Chi Minh spent those first thirty years
watching (with his poet's eye), studying, organizing, and writing in preparation for the real work of his life—those next thirty
long years of struggle to liberate his country.

Back in Vietnam during the Second World War, he allied

himself and the Viet Minh with the United States. The Viet Minh rescued downed American pilots and provided handy information about the Japanese. Ho Chi Minh was known to the OSS as Agent 19. When the French returned after the war and took up where they left off, he led the war until the French defeat in 1954—the war of liberation—and then turned around and fought the Americans. He died in 1969, six years before the final end of the war and the reunification of his country—the focus and ambition of his whole life.

An extraordinary man—like Lincoln, say, who was also a poet.

The morning of Ho's birthday, Larry and I traveled out to his village. We were not the only visitors that day, but we were certainly the only white foreigners, got "points" for that (all day long), and were welcomed with elaborate, if casual, ceremony by the young hostesses dressed in luminous, raspberry silk *ao dais*. The young women all had luminous black hair (gathered at the back of the head with a large, ornate pin) hanging down to here in back. There was much demure smiling, much looking shyly down, many elegant, porch-polite handshakes. Under the delicately sheer, hand-stitched tunics, each young woman wore a T-shirt, I swear, with the word "Hello" embroidered across the chest, Hanoi-fashion; purposeful, no doubt, ironic and charming all at once. Hello, indeed. The houses and grounds were surrounded by kitchen gardens and very-well-tended plots of soybeans and such; a long thatch-roofed residential house; and across the plain dirt yard the open-air schoolhouse where his father taught. The whole, modest compound smelled of ripe, musty thatch. Our young woman guide walked us slowly through the house, laid out one room behind another, shotgun-style, and explained to us in a thin, reedy voice (in practiced and precise, almost drawling English) that *here* was the pallet where Uncle Ho slept (a ballerina's slow gesture of the arm and hand); *this* is the bookshelf (hanging from the low, fragrant rafters) where Uncle Ho kept his several books—the contraption swinging from ropes; *this* is the schoolhouse where his father and some of his friends would sit around in the evening, sipping tea and talking politics—his father, by the way, a respected leader of the village and the countryside round about by virtue of his status as a teacher (the only work he could get). To this day when the teacher enters the classroom, the students rise as one body with greetings, from grammar school to university; such is the automatic status of respect rendered teachers in Vietnam. You could well imagine the boy who would become Ho Chi Minh leaning up against a corner pole in the warm and heavy amber air of the lamp-lit dark, listening with both ears in front as the men gathered in the cool of the evening to read poetry, palaver about history, and discuss the hard facts of French colonial rule; the boy soaking it in with all the passion and zeal and determination of a quick and inquisitive mind.

Back in town that evening there was a birthday party. The word of our visit to Ho Chi Minh's home arrived ahead of us. We were welcomed elaborately, invited to a concert, and obliged to sit at the high table with its flowers and fruit, tea and beer, reserved for the big potatoes. The concert hall was harshly lit with banks of bare fluorescent lights, and very warm. And it was like every bad high school dance I'd ever been to—cheap sound system and fuzzy speakers, a young woman singing bravely, and enthusiastic dancing. Uncle Ho would have been proud. The large windows all around were filled with the faces of young kids who not only couldn't get in, but were clearly not wanted. Two cops stood at the wide front doorway holding cattle prods and would threaten with growls and gestures the kids who tried to scooch in. The cops would make *that* move, and the kids would scoot back. Then the cops would turn back to watch the concert and the dancing and the bopping to the music and the beautiful young woman crooning along, and the kids would scooch up again. The cops would get earnest again, always looking as if they were going to zap somebody (but never did), and the kids would smile and wiggle and scoot back again. This was repeated many times, and each time the kids would get a little further in, spilling into the room like a nine-yard dump truck load of sand, until the cops finally gave up—hey, it's a birthday party. The dancers completely ignored Larry and me, but the band and the singer, who showed up wearing a red leather dress, looked right at us while they played, smiling big.

After the party, Larry and I made our way through the dark to catch our overnight train to Hue. The train was full, and Larry and I had to take bunks in separate compartments. At 5:00 the next morning, just at sunup, the train rolled to a stop out in the middle of nowhere and everyone got off, but no one told us why. It took Larry and me a minute to figure out that everyone on the train was lining up at a large, round well near a railroad outbuilding to "refresh" themselves. For a nickel you were given a pan of fresh-drawn water, a silver of waterlogged soap, a towel about the size of a kitchen rag, and invited to wash; we had about as much time as it took to wash our face and ears (comb the hair later). While we washed up, Larry told me that he hardly got any sleep because of the hooker in the bunk above him, who conducted a brisk business right up until the time we stopped. She would signify to the next customer to come and get it, he said, by lazily stretching her arm out and rubbing her thumb across the tips of her fingers—that universal gesture for "money." And all that night long, he said, neither she nor any of the men ever uttered a peep.

Back on the train, peddlers by the dozens, the scores, walked alongside the cars with baskets under their arms selling "breakfast" through the open windows. Everything from quick gulps of tea to Hong Kong Coca-Cola, bananas and palm-leaf-wrapped patties of sticky rice garnished with an almond, and other "finger food." This was quite a business. For instance, say you live in a town along the line, and half a dozen trains or so roll into town each day and stop for four or six or ten minutes at the most to discharge and take on passengers, swap baggage and mail, and other such things. To make a little extra cash, you get yourself a winnowing basket or scrounge a cooler and walk up and down the train, peddling "stuff" through the windows—always open for the breeze. Altogether you have half an hour, say, to make the nut for the day; a pretty hard dollar.

That was why Larry and I always bought something in the stations, anything, even if we weren't hungry or didn't need it just then; we always shared around with our compartment mates.

Meanwhile, other peddlers mounted the roofs of the cars along with the folks mooching a ride to Hue, the next stop down the line. And after the train moved off, the topside peddlers went to work. These guys always worked in pairs, because they had to. One guy would grab the other's ankles, and he would swing himself over the side and slide along from one compartment window to the next, selling Coke or beer or bananas or cigarettes or what have you; I don't know that I'd ever lived until some guy appeared in our compartment window upside down, long hair a-flying, holding a warm can of Coke rightside-up, pointing to it with his other hand, and urging us to "Buy!" And we did, even though the guy was pushing pipinghot Coca-Cola. Our compartment mates thought we were crazy. I told them that the peddler gets the money, but I get a story. "Oh, well, in that case . . ."

We knew we were coming into Hue when we crossed the Song Huong, the River of Perfume, and glided through the neighborhoods into town.

Of all the cities I've visited in Vietnam, I like Hue the most; a combination of Savannah, Georgia, and Cambridge, Massachusetts. For me, Hanoi, yet and still, has the feeling of the forbidden—funky and busy and somewhat arch—in that way of all national, political capitals. And Ho Chi Minh City—which everyone still calls Saigon, and which simply vibrates (some things die hard)—was, is, and will always be the big-city commercial center of the country, in a style instantly recognizable to any New Yorker. There will always be money coming and going in Ho Chi Minh City. Pearl of the Orient the French called it.

But Hue, a city of about a million souls in the geographic middle of the country, and the imperial city of the last Vietnamese emperor, Bao Dai (who ruled from 1925 to 1932), has

about it an undeniable patina of ease; beginning with our quiet,
easy-does-it pedal from the train station straight up Le Loi
Street along the river to the guest house; hotels and schools and
a hospital on one side, the park and cafés on the other; the
Citadel across the way where a great red-flag-with-star (as big
as my backyard) rolls and furls in slow motion with the breeze.

That afternoon we got the tour, beginning with the famous
Citadel—the walled neighborhood around the emperors'
fortress called The Great Within, where the man and his family
and household lived. In 1968, during the Tet Offensive, the
Citadel was occupied by North Vietnamese troops, who hoisted
the flag of the National Liberation Front, the VC flag, we called
it. I have good friends, ex-Marines, who fought in the Battle of
Hue, skirmishing house to house on their hands and knees up
Highway #1 to the river. A peculiar sensation bubbled up in me
when I stood along the high, thick wall of dark stone and looked
down over the cannon embrasures, across the river to the hospital compound, the imperial school where this or that emperor's sons went to school, and the brushy, formal park
grounds—it made me uneasy, I tell you, because it was clear that
anyone assaulting the fortress was easy pickings (as they say);
first the river, then the grassy park, then the moat covered with
lotus fronds and duckweed, and then the fucking wall.

That evening, our Writers Association host, Vo Que, invited
us to dinner at a place just inside the Citadel. The table was spread
under a large tree hung with paper lamps. Peppery soup, baked
fish, roll-your-own spring rolls (shrimp, cilantro, and slivers of
vegetables), squid stuffed with more shrimp (and dipped in exceedingly hot shrimp sauce), spiced beef (dipped in fish sauce,
called *nuoc mam*), and a kind of satay chicken dipped in salt and

pepper mixed in lime juice; beer and coffee and tea. I never had anything but a wonderful meal in Hue. There is a saying that there are 3000 recipes in Vietnam, and 2,500 of them come from Hue. Dinner over, out came the guitars and one-string violins—the singing and poetry—and more beer. I have to say that Vo Que, born and raised in those parts, is perhaps the happiest man I have ever met, cheerful and engaging with a youthful spirit; poet, singer, and musician, he was most interested in reviving the music of the emperor's court and well known in town. In fact, hanging out with Vo Que was like palling around with Elvis, and everywhere we went everyone seemed to know him, and he seemed to know them; he was even importuned on the street as we pedaled along, touring; watching him, I'd say that he was as good as giddy with pleasure at his renown. During the war he was a student in Saigon, opposed to the government, and did a year in prison for that; after prison he came back to Hue, joined the revolution, went to the woods, and worked as a journalist. In all the conversations we've had, when the talk came around to that year in prison, it was the only time I saw tears well up in his eyes (many died, he said, including children), but then he'd quickly compose himself, grab the guitar, and sing something.

After dinner, we moseyed down to the river, climbed aboard one of those narrow river boats tricked out for parties, accompanied by a four-piece band and several young woman singers. This was a custom begun by the Nguyen kings, taking in the cool of the evening (in the same way that Chicagoans take to the beach of an evening in summer—it's always cooler by the lake). We went up the river a good long way, turned the motor off, and drifted back into town. More tea and beer, singing and poetry and talk as we easy-drifted with the current past the little lights of houses and upside the hills in the dark distance, and all the while overhead great flashes of heat lightning banging through the monsoon thunderheads above us.

The next morning bright and early, Vo Que arrived at our guest house followed by a cloud of dry dust. "Hie-na-man! Rote-man," he shouts. It was time for our obligatory river tour; the Tien Mu Pagoda and Emperor Ming Mang's tomb, upriver. We board our boat, pack a cooler of beer and Cokes, swing out into the river, and make our way upstream past the Citadel, under the railway bridge, and out of town. We slowly pass similar boats loaded to the very gunnels with fresh gravel and sand hand-dredged from the river bottom to be sold for construction—highway roadbed and makings for concrete and mortar; there is construction everywhere—houses, business, and hotels. It was clear that if you worked the building trades, bricklaying especially, you would never want for work.

Chua Tien Mu (Heavenly Lady Pagoda) is a working monastery where Buddhist monk Thich Quang Duc lived; Quang Duc is famous to the world for the dramatic manner of his death. Hue has always had a reputation for its activist Buddhist monks; the monastery where Thich Nhat Hanh, now living in France, studied and taught is just across the river. In 1963 the Buddhists were so fed up with President Ngo Dinh Diem's religious suppression that they began organizing all over the south. The marches and rallies were all the more strange because the banners carried by the saffron-robed monks were written in plain English so that the American press and the American people could not mistake the message. One day Quang Duc and several other monks piled into a vintage robin's-egg blue British Austin four-door and drove to Saigon; and judging by the river road into town, the trip from Hue to Saigon must have taken forever. They drove to the protest site; Quang Duc got out of the car and assumed the lotus position of meditation in the middle of an intersection; his companions fetched a couple cans of gasoline from the trunk and poured it all over him; and he struck the match and set himself on fire. Shortly before, Malcolm Browne of the *New York Times* received a phone call that something noteworthy was going to happen at the protest march, and so he was the only Western journalist there and took the photograph that shocked and outraged the world, of Quang Duc sitting cross-legged with his hands in his lap, going up in flames. He was the first of many such self-immolated Buddhists—there would over the next several years be enough to fill the calendar. To give you an idea of how little the Diem government thought of the Buddhists and their protests, when asked about the incident, Madame Nhu, President Diem's uppity sister-in-law, snorted and scoffed and called the protest "barbecue."

The robin's-egg blue Austin, now streaked with rust, is displayed as a venerated object across the way from a large garden of bonsai behind the pagoda; in the windshield is a copy of Malcolm Browne's photograph. The story goes that when Quang Duc's body was properly cremated—as is the custom—his heart would not burn. Three times the monks tried to cremate it, and three times it would not burn. This was, of course, taken as a holy sign, so his heart was put aside; his name is always spoken with great reverence.

Back in the boat, we take off upstream for the tomb of Ming Mang, the second of the Nguyen emperors, who ran the country from 1820 to 1841. While the tombs were being built, Emperor Mang had the habit of taking his long weekends there, where he would, without a doubt, party down. The tomb compound had more of the look of a park about it—party spot was perhaps more to the point—and there was little of the courtly martial air that we saw at the others. Emperor Ming Mang had, by the way, 40 wives and fathered 150 children, and was by all accounts (and the abundant evidence) a randy old goat, though he didn't live to see fifty. These days he is remembered for the liquor that bears his name, a whiskey concoction that is supposed to improve a man's "capacity" for love. At the tomb we stopped for lunch, a couple rounds of the emperor's whiskey, and more singing. We took turns pouring everyone a shot into everyone's teacup, and then hammered them back. We toasted Emperor Ming Mang many times, and I told Vo Que and the other poets that Ming Mang proved that it was *good* to be king (as the poet once said). I've never visited Hue but that someone hasn't presented me with a bottle of good old Ming Mang—about as medicinal a whiskey as you could ever imagine; the way to drink it is to take a little in the mouth, swish it around to deaden your tongue, then knock the shot right back—first of the day.

On the way back to town, the only guy on the boat not sleeping was the boatman, and about halfway to town he pulled to shore under the shade of a bamboo hedge (the stalks about as big around as your leg) and asked, since it's so godawful hot, if anyone would like to take a swim. Neither Larry nor I, of course, brought swimsuits. I asked about the boatman's wife, but she wasn't concerned (not after all the Ming Mang jokes), and, grinning big (that no-problem, what-else-is-new look on her face), motioned for us to go ahead—she definitely had better things to do than gawk at us. There was a brief discussion about pollution of one kind or another, and then I said what the hell, I will always be able to brag to my grandchildren that I swam in the River of Perfume. We shucked down to our Jockeys, asking just how deep it was here. It's fine, everyone said, and cool on such a hot, hot, hot afternoon. I laid my clothes over a chair, stood on the gunnel at the bow, dove in with a leap, and drifted underwater a good way into the current; ah.

The next morning we caught the train to Ho Chi Minh City, and almost immediately I learned one of the great lessons of second-class travel aboard the Vietnam Railway; always hang out in the club car—or what passes for one. Soon after the train took off south, Larry and I went looking for it, laid in just behind the engine. The car was a converted coach car with the seats ripped out and replaced with booths; this was where the crew hung out. Toward the front of the car, cases of beer bottles were stacked to the ceiling along with sacks of rice, bags of vegetables, and crates of chickens. In the door forward were two stone braziers topped with two large woks; two of the women on the crew did all the cooking for the folks who had bought the meal plan. We sat and asked for a pot of tea, and almost immediately one of the youngsters came over and started in on his English. He was Brakeman #3, and said that the conductor—there—was inviting us to join him for lunch. So first before anything, the two cooks whipped up some lunch for the crew. We were veterans, the conductor and several others were also veterans, and we ate and drank and talked among ourselves as well as Brakeman #3's English would allow. I fetched one of the bottles of Ming Mang that Vo Que and the Hue poets insisted we take along—we were after all, they said, going to Saigon and would need it. I presented the bottle to the conductor. Brakeman #3 kept whispering that the conductor was "a communist," and we kept telling Brakeman #3 that the war was over, had he not heard? After many rounds of beers, several shots of the miracle elixir Ming Mang, toasts, and much rambling conversation, the conductor stood up, twirled his watch loose around his wrist, and announced to one and all that it's nap time. And so it was; I was getting the feeling that I have only one liver to give for my country.

The next morning we rolled into Ho Chi Minh City, and we could not get to our hotel quick enough. We need a shower, we need some air-conditioning; I need a nap.

5

The Black Virgin Mountain

So. The last day of our trip Larry and I hire a car. Our plan is to drive north along Highway #1 to Cu Chi, then drive on to Trang Bang, hang a right on Highway #22, and move on to Tay Ninh City, the Cao Dai Temple, and the Black Virgin Mountain; what the French called La Montagne de la dame noire, and the Vietnamese know as Nui Ba Den.

We catch an early breakfast at our hotel, around the corner and down the street from city hall, the old Caravel Hotel, and the good old Rex. The Rex Hotel, by the way, is famous for its wartime rooftop bar where the big-potato lifers and white-collar spooks, the world's journalists and get-rich-quick Saigonese entrepreneurs gathered to watch the war. They'd sip their cocktail-hour aperitifs and nightcap whiskeys and jaw among themselves about how well or badly "things" were going. Many a prize-winning American talking head bronzed his reputation, once and for all, in Southeast Asia by watching the parade, down *there*. The story goes that when the North Vietnamese cadre waltzed into the hotel to announce the hotel's closure as a symbol of French and American oppression, the maitre d' made the case that what few foreigners would travel to Saigon should be allowed what hospitality the hotel could provide; the hotel stayed open (though the amenities were considerably reduced). By the 1990s the doormen are back, ready to open the door for you, and the bellhop is ready to escort you to the elevator, the barmen on the roof bring your drinks with a snap and a smile, and the Rex is back in business.

I know that Saigon is a beautiful city, but I have never been much of a Saigon person. In March 1967, when I arrived as a pasty-faced fucking new guy in my closely tailored Class-A khakis, we were taken straight from Tan Son Nhut to the 90th Replacement Depot in Bien Hoa aboard cranky old olive drab school buses with chicken wire fitted across the windows to prevent someone from tossing hand grenades at us. At the 90th Repple Depple, we waited in our hundreds for our in-country orders, which could take days. Meanwhile, we were kept busy with work details of one kind or another. I swear on the heads of my children that the very first work detail I was assigned to was burning shit, but without the saving grace of sharing a joint around while watching the sun go down in a blaze of glory, listening to Grace Slick belt out "White Rabbit." In Vietnam the latrines were simple wood-and-screen huts with half a dozen and more "shitters." Underneath each hole was half of a fifty-five-gallon drum, and, of course, clouds of flies. You slid the shit-filled drum-halves out, lined them up, poured diesel fuel in each to the brim, and then lit them. The diesel smell was bad enough, the shit smell was bad enough, but when you put the two together it produced a gagging stink that caught in your throat and which is difficult to describe. You lit these concoctions with your Zippo, which often took some doing because diesel doesn't just "catch fire" like gasoline, and then you stood there in the midst of this brand-new smell of greasy diesel smoke, shit, and piss, leaning on your ditch spade to make sure the fire kept going. Welcome to Vietnam, cousin. I couldn't get out of Saigon fast enough, and I didn't see the city until a year later, going home, when I arrived back at the 90th Repple Depple in the afternoon, spent the evening drinking beer with a bunch of homebound Australians and half the night sitting on the tin roof of our hooch smoking grass while we watched the occasional 122mm rocket coming out the near woods, screaming in a low arc across the sky, and making its way toward the city—glimpsing the far-off flashes of the explosions and listening for the thud; a large, clean sound. The next morning we dressed in our year-old, ill-fitting khaki Class-A uniforms, gathered our paperwork, were transported in the same buses with the chicken-wire windows, and spent the rest of the day waiting for our plane. My uniform all but hung on my body; after all, when I arrived I weighed 160 pounds, and was in the best shape of my life. The day I left I weighed something like 135. We smoked some more grass to assuage the sharp anxiety of the absolutely bitter end of the counting and waiting, because we were unarmed and in the open, within an ace of going home, and the Tet Offensive was still roaring right along outside the wire; get me the fuck out of here. That's when I started on the pocketful of downers the medic had given me back at Dau Tieng.

When our plane came, the fresh load of fucking new guys got off, we loaded up, and in no time at all we were gone, gone, gone—and everyone exhaled.

No, I can't say that I was ever much of a Saigon kind of guy.

Around the corner and down the street from the Rex Hotel, our driver fetches us in our hotel dining room. He's parked at the curb, let's go, and both Larry and I are more than ready to go; even eager. After all, it's the Black Virgin Mountain we're talking about.

The kid's in his early twenties, dressed in dark pants and white shirt, black shoes and white socks, and we start right in calling him "Ferguson," after that character in Mark Twain's *The Innocents Abroad*. There is a relish in his smile, a jaunt to his walk, a ripe and pleasant swing to his whole body. He has the bright, scrubbed look of a guy right out of chauffeurs' school, in love with his job, and today he's going to haul a couple of middle-aged, round-eyed white guys up to Tay Ninh and back. He will drop us off at Tan Son Nhut Airport around ten o'clock for our midnight flight back to Bangkok.

He will have us to himself all day. What a lark.

We leave the hotel and drive straight into the very teeth of the rattle-bang-whoosh, go-getter Saigon morning rush hour, everyone going every which way. From first light the streets have been chaos, and getting to Highway #1 takes forever. Main-drag street markets and ma & pa sidewalk noodle shops, cafés, bicycles and dawdling foot traffic, one-way streets and miniature traffic lights, and the whippy buzz of knee-to-knee scooter traffic that doesn't pay any mind to anyone's horn but his or her own.

Like the best big-city drivers everywhere in the world, Ferguson knows all the cabdriver shortcuts. Slowly, but surely, we make our way to Highway #1, which comes over the river from Bien Hoa, then zigs and zags its way through Saigon much the same as old US Business Route 66 wiggled its way through Dallas.

Highway #1 takes you northwest out of the city to Hoc Man, Cu Chi, Trang Bang, then Go Dau Ha and on to Cambodia (that sad, sad place) and Phnom Penh.

We drive with the windows open through every raw and fragrant aroma you can imagine. French-built Old World street sewers, market middens, winnowing baskets piled with catch-of-the-day fish, bamboo racks of fly-blown pork joints, rough mahogany trays of fresh-baked bread, head-high piles of rice in hundred-pound bags, wood fires and cooking smells, and the heavy saturated industrial stink of turpentine, spray lacquer, and the thick ozone smell of acetylene from a cyclo factory.

Here we are, two American ex-GIs all but joyriding to Tay Ninh City to see the Cao Dai Temple and the Nui Ba Den. And I don't know about Larry, but I feel anything but blasé. It's like touching lightly that tight itch in the stitches back of your arm; it's the sweet, metallic toothache pain you rub with the tip of your tongue again and again; that moment in a drowning dream when your lungs are all but bursting and you finally give up and take a breath, but nothing much happens except your lungs fill with water, and you keep swimming.

All morning we pop tape after tape into the dashboard player, listening to virtuoso Ozark banjo and mandolin, mellow and spirited Miles Davis, raw and yappy Rolling Stones, and get-down-and-get-some wang-dang-doodle Chicago blues; American music that Ferguson has never heard. He chimes in with Vietnamese love songs and lullabies. Larry and I respond with "This Land Is Your Land (This Land Is My Land)," "The City of New Orleans," and the ribald version of "She'll Be Comin' Round the Mountain."

We make our way to Highway #1, and head north—still bucking the morning rush.

We pass Tan Son Nhut and the far neighborhoods of the city. Soon enough we're out of the city, and the traffic takes on the character of over-the-road commerce: intercity buses headed for the countryside piled with boxes and bikes, crates and cloth-wrapped luggage, and crammed with passengers; nine-yard dump trucks hauling oven-fresh bricks, bags of sand, and lime; ex-GI diesel trucks rigged out to haul fresh-cut timber, and Army jeeps tricked out as short-haul jitneys.

We pass the nearly invisible shin-high, lozenge-shaped red-and-white milestone for Hoc Man, and then we're cruising through ordinary country life. Along with the buses and trucks and jeeps, now there are bicycles and ambling foot-traffic, loafers and little old ladies scooting along hauling buckets full of something hung from sagging shoulder poles. Ferguson works his way past oxcarts, herds of cattle and large, sleek water buffalo, gaggles of geese and ducks; roadside village markets, *pho* shops and cafés, bamboo hedges and banana groves, broad ditches and wide stretches of farmland, and far-off wood lines.

Then in the middle of all that wide, green country we encounter, inexplicably, a parade of young men leading small horses toward the city; by twos and threes the casual parade stretches a mile and more. These are sleek thoroughbred horses; petite, like polo ponies. The horses prance and shine, fawning proudly like all young and beautiful things. The spectacle is explained easily enough without knowing for certain: in Saigon there is a racetrack, built by the French, of course, and it is apparently still holding races (*some*body in town has serious clout). Vietnam is most assuredly a socialist country, and horse racing—the literal, proverbial sport of kings—is, you should not doubt, very much officially frowned upon, but these are definitely *not* draft animals; it would be a crime against nature to strap them to a cart or work them in the fields. Where else could they be headed but the racetrack? The grooms walk along easy and flat-footed (the way those guys do) in the dry dust and gathering oppression of morning heat, and the horses, diminutive and delicate, almost miniature, all but dance with downright dainty steps and eager shakes of the head; that fresh, bright look of well-honed athletes in their eyes (as if they are as eager for a day's racing as anyone).

Inexplicable, indeed, like the Mozart sonata we heard one afternoon coming from the back room of a T-shirt shop in old Hanoi (played by a ten-year-old on an elderly, out-of-tune, saloon-parlor upright piano); like the guy we saw on our Hanoi–to–Ho Chi Minh train handcuffed within an inch of his life to a coach seat—an entire car to himself, except for the heavy canvas mailbags crammed to the rafters (a prison-bound convict, the conductor told us, who has been hard-traveling for a couple days, will hard-travel some more, and then will travel no more); like the guy in Haiphong taking his extreme ease in a doorway with a baseball cap pulled over his eyes (across the front the caption read: I Know Jack Shit, the hat worn with such casual aplomb that you just *knew* that somewhere in the world was a man with the unfortunate name of Jack Shit, and this kid knew him right well). As inexplicable, I say, as the billboard near the

Hue train station that read in large letters and in plain English, AIDS KILLS; like the village we passed on the train south of Vinh where one of the market street open-air shops had a sign over the front that read HONDA GUY; like the packs of Harley-Davidson brand cigarettes at the Da Nang train station.

Past Hoc Man is the milestone for Cu Chi, just up the way. The town still straddles Highway #1 and is still profoundly undistinguished.

During the war, Cu Chi was headquarters for the 25th Infantry Division, nicknamed "Tropic Lightning"; the Pineapple Division of the "old army" at Schofield Barracks in Hawaii, of which James Jones spoke so eloquently. The shoulder patch—a bright red taro leaf with a chop of yellow lightning down the center—was known as the "electric strawberry." Well, the mighty and terrible 25th was "tropical" all right, but there was never any "lightning" to it that I ever saw.

Rest assured, all these years later no one remembers Cu Chi for the American base camp (long ago dismantled board by board and carted off; vanished as if it had never existed), but rather for the tunnels that spread out beneath us for two hundred kilometers—all the way to Saigon, some said. Abandoned after the defeat of the French, the tunnels were resurrected in 1959 and for the next ten years much expanded. This was the "thing" I'd spent my whole, long war-year looking for, more or less, so this trip I figured the least I could do was find out what the tunnels actually looked like; sheer, rhetorical curiosity.

The Vietnamese, to no one's surprise, have turned a portion of the Cu Chi Tunnels into a must-see stop on the "war

nostalgia tour"—gift shop, rifle range, and all; think the backyard version of Frontierland at Disney World; think Boxcar
Willie's in Branson, Missouri; think Hunter Thompson's Las
Vegas. I don't know that the Vietnamese go near the place, except for those obligatory busloads of grammar school day-
trippers. The tunnels have to be one of the great running gags
of the war as well as one of the great puzzling myths—a vast Gor-
dian complex, downright cleverly put together, though I don't
know that the Vietnamese think of it that way. And we knew the
tunnels were there, the sons-a-bitches were hardly a secret,
but we just couldn't fucking find them; not with bombs, not
with bulldozers, not on our hands and knees; the institutional
clumsiness of large, overequipped, and spectacularly arrogant
armies from postindustrial countries is well documented.

We roll into Cu Chi and Ferguson turns off Highway #1, follows the farm road out of town a piece, and turns down a cart
trail that looks for all the world like a tractor path. Even though
Larry and I both did time at Cu Chi, we see nothing that looks
vaguely familiar—why would it?—so the actual site of the base
camp could have been anywhere. In the brushy scrub to the
side is a "crashed" Huey chopper; it looks as if it's been shoved
off the back of a truck; without a doubt part of the *ambiance de
guerre*. We soon arrive at the tunnels proper, and Ferguson
pulls into the wide parking lot, raising clouds of choking, dry
laterite dust, and parks in the shade; the heat has gathered for
the day, and it will definitely get hotter still.

For a modest entrance fee we are welcomed into the park
and "the tour."

Ferguson has absolutely no interest whatsoever in the tunnels and much prefers to hang out in the café, hitting on the

waitresses and gift shop girls with his Saigon city-kid jive, and wishes us well. As we walk away, he takes a fluffy, long-handled feather duster out of the trunk and wipes road dust from the car with long, leisurely sweeps of his arm.

On the hour, the visitors are ushered into a thatch-roofed open-air hooch and invited to sit among the double rows of rough-cut homemade benches arranged lecture-hall fashion. The dozen tourists sit quietly; several other carloads of European tourists, young Scandinavians and Germans on vacation (the word is spreading that Vietnam is cheap travel). One of the young women leans at us and asks, "You are soldiers American?" *Ex*-soldiers, we tell her and the rest. "Oh," she says (with all those stories of rape, pillage, and murder by the Americans rattling around in her head). The Scandinavians and Germans buzz about that. Finally a uniformed park ranger, obviously a big potato, walks in under the roof, stands in front, asks for everyone's attention with a voice of heavily accented English, and a moment later we are all ears. Behind him is a permanent "visual aid" about the size and heft of a four-by-eight-foot slab of plywood— a large, detailed illustration of the tunnels rendered in bright-lacquer, road-sign colors. Across the top it reads: The Tunnels of Cu Chi; room enough, you may well imagine, for ten thousand souls. Barracks and training facilities, kitchens and latrines, booby traps and deadfalls, main drags and dead ends, munitions factories and weapons caches, ammunition dumps and firing ranges, rallying points and sally ports, spider holes and sniper positions, knee-high log-and-dirt bunkers complete with embrasures. Here and there little bitty guys crawling along on hands and knees; crouching at the firing step of a spider hole; sitting with their heads together discussing something very important; the litter wounded recuperating just fine and dandy; everyone just as busy as busy can be. Aboveground, clusters of little bitty Americans stand in the full light of day and stare off into space, cartoonishly oblivious. If you didn't know any better you would think you were looking at a mail-order ant farm.

To one side of the lecture hall is the other permanent exhibition: a homemade gun rack with AK-47s, M-16s, M-79 grenade launchers, pistols, small-bore 60mm mortars, RPGs, and other "vintage" hardware; weapons I hadn't seen close up since before the end of the war—the guns and such either slathered with oil or rimed with rust.

In the other direction, the young tour-guide park rangers lounge in low-slung hammocks set out among a grove of trees, taking a break with that universal expression of dumb-job boredom.

The big-potato park ranger stands in front of the "Tunnels of Cu Chi" and begins his boilerplated lecture. He, of course, knows the thing by heart and gives it to us with that same droning blah-blah-blah you get when you visit Colonial Williamsburg, the Kennedy Space Center, or the lately spiffed-up immigrant repple depple on Ellis Island in New York Harbor. Think Sutter's Mill where California gold was first discovered (demonstrations of finding the fist-sized nuggets lying on the ground); think New Salem, Illinois, where Abraham Lincoln grew to manhood (demonstrations of splitting wood rails with one swing of an ax, walking a hundred miles to return a nickel, and educating oneself by reading Shakespeare and the King James Bible by the light of a fireplace).

Our brave fighters, the guy says, began digging the Tunnels

of Cu Chi in 1959, taking up where they left off during The Liberation from those dogs, the French. True.

Our brave fighters built the Tunnels of Cu Chi with ordinary garden tools. True.

The American imperialists of the 25th Division built their main base camp over a portion of the Tunnels of Cu Chi and were never able to find them. Only too true.

Here—and he whacks the illustration with an arm's-length pointer of delicate bamboo—here, he says, our brave fighters did *this*, here our brave fighters did *that*, here our brave fighters did this *other thing*.

I cannot help but think of Gen. William Westmoreland's over-the-top, dumbjohn "light at the end of the tunnel" crack, one of those classic, dufus remarks that give definition to the hopelessness of the military mind; and Michael Herr's repeating it to a 25th Division "tunnel rat," and the guy looking up and snorting, saying—rightly enough—"What does *that* asshole know about tunnels?" Suddenly the simple awareness of the park ranger's deadening presentation overwhelms me. Right then and there—the weapons, the dumbed-down lecture rendered with a straight, lifer's face, the guy dressed in rumpled fatigues hung with all his "stuff"—that moment, *suddenly*, an uncomfortable disquiet gathers in me, rises (like a wiggly bug crawling up the back); a tingling from the waist down, as if half my body has gone to sleep. And all at once, I sincerely, profoundly, do not want to be there. Then it occurs to me that the Tunnels of Cu Chi *must* be an army installation; these *must* be military cadre; no wonder the droning boredom would fell an ox.

The Vietnamese are pretty quick about everything else, what happened here?

At Ho Chi Minh's childhood home near Vinh, for instance, the good city fathers had the common sense to assign drop-dead beautiful young women to lead the tours; this in the same canny spirit that the Hall of Presidents at Disney World is staffed by attractive, neat young "hostesses." Whether the twenty-five-words-or-less boilerplated lecture is the gobbledygook bullshit of outright propaganda, the idealized and wooden good intention of national myth, or the real fucking article, the tour is automatically guaranteed respectful attention because an attractive, well-spoken, well-mannered young woman is reciting the thing with an all-too-palpable reverence in her voice (softly sweet so that you *have* to pay attention); the poised, porch-polite body English and perfect hair; and the moist twinkle of naïve, intense sincerity in her eyes that make the mothers in the group "so proud" and the fathers, well, attentive with fatherly admiration.

But there is none of that at "The Tunnels of Cu Chi."

I have been told that every once in a while a middle-aged Vietnamese woman quietly attends these presentations and takes a seat in back. During the war she had been a sniper (so the story goes), and at some point in the guy's palaver she would stand up and talk of the hundred and some Americans she shot during the war. You can imagine her rising from her seat with the full righteous dignity of exaggerated patriotic pride, pointing straight to one of the illustrated spider holes. "There!" she will say. "I was a sniper for the Revolution, and stood *there!*" and raises her arms as if holding a rifle to her shoulder, squinting down the sight, and laying her finger to the trigger. Then she will squeeze, whisper *"pow,"* and her body will twitch at the recoil. Then the rifle will disappear when she opens her arms, again pointing. "I was a Cu Chi tunnel rat. I stood there. I shot *many* American imperialists."

Moving on, when the lecture-demonstration is nearly finished, the young tour-guide park rangers rouse—they *also* know the cadre's speech by heart—and by the time the lecture concludes they are standing along the edge of the hooch shade, motioning for us to follow them.

I am only too glad to get on my feet and move.

Every second person is given a flashlight, and we are escorted through the hammock grove to a narrow footpath cut into the thick, scrubby woods, where fifty paces or so down the way we come to a bit of a clearing the size of a small bedroom. The ground is all but bare, and the ranger invites us to find the tunnel entrance. "You are standing within two meters of it," he says. The Scandinavians and Germans look at us, as if Larry and I have x-ray vision. After a very long and polite, rhetorical moment, the kid reaches past Larry and brushes aside a scatter of leaves, and reveals a tiny loop of ordinary household wire and a small square—as if someone had scratched a rectangle in the dirt with a stick. He slips his thumb through the loop, gives it a yank with the flip of his wrist, and up pops the cover—a square, tin thing (like the shallow oven-tray for cafeteria fruit pie), filled with hard laterite clay and topped with loose jungle trash.

Lo and behold, ladies and gentlemen, the "Tunnels of Cu Chi."

Moving on, the young ranger pops down into the entrance, invites everyone to follow him, and disappears into the tunnel. Larry and I go next, because I'm thinking first in, first out. One by one the members of the tour make their way, feet first, into the hole; there is no elegant way to do it, and, once inside, there is just enough room to turn around. You squat on your haunches, and there before you is the tunnel that takes you to "headquarters." The tunnel has been generously enlarged, scoured out, to accommodate the hefty, middle-aged American tourists, like Larry and me. Larry is over six feet tall and has been athletic all his life—he is not small. I have never heard him speak of tunnel-ratting during his war-year, and as skinny as I was during the war, I was *still* too big. No, it was always the platoon shrimps that got stuck tunnel-ratting. With the flash-light held in front, you duckwalk (head down) perhaps twenty feet; the person in front of you a vague silhouette. One by one we emerge into a dirt-roofed bunker with skinny embrasures all round (enough, perhaps, for a rifle muzzle), a clean dirt floor and "square" walls, and a conference table of rough-cut timber and plain plank benches all round. The "room" is any-thing but spacious, and if the logs and dirt stand knee-high aboveground, *that* is generous.

In such a place, we are told, the brave fighters of the mili-tary branch of the National Liberation Front planned the war; we could well be directly under "Tropic Lightning's" Cu Chi base camp for all we know.

Moving on, we again squat down and duckwalk into the next bit of tunnel, about the size of an Irish country-cottage fireplace. The tunnel is perhaps twenty-five or thirty meters long; the atmosphere is close and hot. The floor and walls are clammy and sticky, the air rank and stifling, the flat walls and rounded ceiling have been carved out with the sweeping, scal-loped strokes of a gardener's trowel. Your clothes and the back of your head brush against the dirt and roots; the dirt underfoot moist enough to make a sound. You are middle-aged, fat, and

out of shape. You are quickly pouring sweat, quickly claustro-
phobic, and ready to leave; now. Not halfway to the end your
legs and back ache terribly. You stumble on your haunches,
walking—no, staggering—along like a duck. Your only consola-
tion is the soft echoes of the Scandinavians and Germans
struggling along and jabbering among themselves.

This is the first time I had ever been in any of the tunnels.
During my war-year it had never remotely occurred to me to
venture into one—aboveground was exciting enough. I had
never been handed a flashlight and a .45 and *told* to go.

Every once in a great long while we would get lucky and
stumble onto something, but *that* was always fools' luck.

Well, what do you know? A tunnel—my, my; the very thing
we're looking for.

You cannot turn your back and drive away, so the platoon
draws itself into a tight laager and shakes out "skirmishers"
and listening posts. By twos and threes the men spread out
and *look* for another spider hole, a bunker, an airhole, a fresh
bit of dirt; anything. The fucking new guys stare at the hole in
the ground as if we have come upon the dragon's fabulous
hoard.

The tunnel is probably booby-trapped, so—first things
first—you toss in a grenade or two and stand aside.

You never know. The guy down there just may throw it back.

You wait those four and one-half seconds for them to
blow—*shoong! shoong!*

You listen for shrill screams of bloody murder; a sound like
no other in the world. Instead, a foul, dead-gray smoke rises

from the hole like the last, thick wisps of smolder from a doused leaf fire. It is time to tunnel.

You put the rest of your grenades away. A hand grenade, now, is suicide.

You take a long drink of water; you may be at this a while. Everyone wishes you luck. The lieutenant tells you to be careful.

You promise.

You never wear a shirt, so you don't have to strip to the waist.

You zip your flak jacket to your chin, chamber a round in the .45 (clicking the safety off), snap the flashlight on, get on your belly, and lean headfirst with a long stretch of your whole body down into the tunnel entrance (someone holding your ankles).

You pause to accustom your eyes to an atmosphere of dark and smoke; alert, instead, for the smell of shit; of blood.

You swirl the light around—*looking*; for sandal straps, for scraps of clothes, for body parts, for a splash of eviscerated slime; for—you don't know what.

You *listen* with both ears in front; the brush of a trouser leg; the swallowed groan in a throat; the oiled slide of a rifle bolt; the slip of a hand on dry dirt; the click of a belt. Nothing.

You ignore the sting of smoke in your eyes.

You eyeball just there for the threads of tripwires pulled taut. All you see are the scuffs in the dirt where the grenades blew.

You shinny down and gingerly move into the tunnel no bigger around than a Thanksgiving turkey platter. The air is hot and rank and lingers foul on your tongue—all of Earth is close around.

You lie on your belly, holding yourself up on your elbows as well as you are able. Straight down the way, the beam of your flashlight disappears into an abyss of grenade smoke and the dread of serious dark.

So, you begin.

You squeeze along on your stomach, low-crawling in exquisite slow motion, pushing with the toes of your boots and the points of your hips, pulling with your elbows; ever so sly.

You are scared shitless—there is, flat-out, no other word for it—but unlike that nitwit Hollywood war movie myth of soldiers shitting their pants in those extraordinary moments of stone-high dread, *that* is not going to happen, because the eye of your ass is puckered-up as tight as a bolt. Your bowels are not going anywhere.

You will die first.

You pause and listen more; feeling along the dirt with a light touch of your hand as if caressing a peach. Nothing, so far.

You stare into that smoky, perfect darkness with unique, perfect concentration. If someone is waiting for you just yonder in the deepest darkness you could ever imagine, you are an easy target because your silhouette blocks the little light behind you; not to mention the flashlight in your hand. He has an AK-47 and thirty rounds pointed right at your head. The flames of the muzzle flashes will be the last things you see in this life. But then, so is he an easy target. Your little light will reveal the tiniest shine of his eyes—just as intent as yours. It will reveal that touch of sweat on his face—just as greasy as yours. It will reveal the small, perfect "o" of his rifle muzzle.

You have a .45-caliber semiautomatic pistol with one round in the chamber (a literal hunk of lead) and seven more in the magazine, plus another eight-round magazine held in your teeth as well as that Filipino bowie knife almost as long as your forearm strapped to your flak jacket. Your pistol is pointed straight at him. At this range, in this little space, even the ricochets will carom and hit.

You are fucked; he is fucked. The world has come to this.

You pour sweat. The dirt clings to your arms; dirt sprinkles into your hair; your eyes. Everything clings.

You feel with your fingers in the dirt; mines, booby traps, and deadfall triggers. It is like drawing your fingers through the dry motes of dust under a bed, touching for coin. So far, so good.

You inch along; move and pause.

You watch and listen and slide your touch; and move again.

You shove along, one little bit by the next little bit; creeping closer to the very center of the cleanest, simplest, most uncomplicated dread known well to every prey. With every scoot forward you leave something of the best of yourself behind. Then.

You *sense* rather than see or touch some thing. It is a shiver in the very air, not unlike that sharp, cold pique of awareness that washes through you when someone is watching.

You lie still, so still that you feel the rhythmic jolts of blood under your arms and the side of your neck.

You slide the heel of your hand along the wall of dirt, reaching, until you suddenly reach into air. Well, you will be goddamned. A side tunnel. Now what?

You draw near by inches and listen. Not a sound.

You sniff the air with a long pull of breath—ah. For that instant, it is like a meditation.

You have never been so alone in so dark a place, so godawful afraid in all your born days. For you, this may well be the last of earth.

You stretch forward.

You hold your breath and turn the flashlight, the pistol, your gaze down the way.

You see and hear nothing. This is very good.

You pause a good long while, collecting your wits, but paradoxically you are as near to tranquil as you can ever recall. Absolutely nothing in this world will compel you to turn and scrutinize your discovery.

You move quietly on.

You stop.

You listen.

You watch.

You do this many, many times.

You push along inch by inch, leaving the second tunnel behind. The air comes cleaner.

You search the dirt all around with your fingers, your light, that most perfect of the keenest gaze.

You sniff the air for gun oil, food, backpack canvas, body odor, anything that does not smell like dirt and smoke and darkness. It would seem, still, the only person down here is you. This is very good, indeed.

You squeeze along, mindful and careful. The tunnel doglegs left. Slow and steady. Easy does it. The tunnel doglegs right. Then.

You do actually, finally, inexorably see some thing. Your breath catches in your throat. Hut! Jesus H. fucking Christ. Hut!

You feel as if you will have a heart attack. Stranger things have happened. The "thing" appears a long way off, like an aura, like an elongated bull's-eye. Think Pollack. Think Turner. Think Caves of Dordogne. Think the ghastly avidity of death's breath and nightmares. The thing looms large in the distance, like a dusky apparition. It is the largest, most vague object you have ever seen.

You set down the pistol ever so carefully and press your palm into the dirt to dry the sweat.

You take your sweet fucking time, because you have got all fucking day.

You turn off the flashlight; no need.

You pick up the pistol. The grip is gritty and sweaty, still.

You move, stretching your body long to make as small a target as is possible in so small a place. Every little bit helps, and, well, you never know.

You curl slowly around another turn. As you move, that thing, that aura, that elongated bull's-eye, becomes the faintest, roughest spread of light. It *is* light. It is the light of men's lives. It is the ambient light from another spider hole. The light of your life. Ah.

You are, however, not out of the woods just yet. Let us not get ahead of ourselves.

You crawl to the end of the tunnel with the same special care with which you entered. Several paces from the end you take a good, deep breath and give a shout.

You do this so your friends aboveground do not kill you with *their* grenades.

You come to the spider hole, climb up using the toeholds, and stumble into the fierce, brilliant morning air.

You emerge covered with dirt and sweat. Your filthy greasy

trousers cling to your legs. Your flak jacket is plastered to your back.

You snap your pistol on safe.

You are back among the two-leggeds, as the Indians would say, and standing in the clean, hot light of day. For a long moment you cannot see a thing.

You are hot, your mouth is dry, and you are within an ace of heat exhaustion.

You drink an entire canteen of water, then another.

You would prefer a good stiff drink, but some days you have to take what you can get. A moment later you are pouring sweat all over again.

You tell the lieutenant about the side tunnel. No, you tell him and the demo guys, you have no idea where it goes. One tunnel at a time, sir. They know exactly what you mean.

You have time to wash up while the demo guys and the lieutenant blow the tunnel, once and for all. Then the platoon mounts up and drives away.

You spend the rest of the day catching your breath. Later that night you smoke marijuana until you cannot see straight.

You drink any kind of hard liquor you can get your hands on.

You go to sleep that night stoned; drunk.

You do not dream. It is not the first of such evenings, and it will not be your last.

Soon enough Larry and I have had more than enough of the "tunnel," but the end is many, many paces yonder; it seems endless.

Moving on, one by one the tour comes to the end, and, one by one, we stand up in the hole and pull ourselves out—Larry and I turning around to lend a hand with the others. The Scandinavians and Germans look like they've had it, too. It is clear that they are really suffering in the heat. Drink some water *now*, we tell them. The tour has all but circled the parking lot. We thank our young Vietnamese guide, hand him back his flashlights, and wash our hands and faces at a long bamboo contraption set up for the visitors to "refresh" themselves. The young ranger reminds us with a bright salesman's enthusiasm that just down this way is the rifle range, where, for a dollar a bullet, we can fire an authentic, war-vintage AK-47; it is not to be doubted that "The Tunnels of Cu Chi Rifle Range" is a real moneymaker.

Larry and I laugh out loud and politely decline with many thanks. On our way back to the car we pass by the gift shop. Inside is rack after rack of T-shirts that read:

I SURVIVED
THE
CU CHI TUNNELS

These, along with stacks of phony Zippo lighters and the ubiquitous bowls of really phony GI dog tags. The Zippos are especially pathetic because not only are they obvious Hong Kong knockoffs, but the engraved drawings are nowhere near pornographic enough nor the inscriptions obscene enough—not from what I recall, at any rate. And the handfuls of dog tags are just plain sad.

Who buys this stuff? Well, apparently Scandinavian and German tourists.

*   *   *

Ferguson is waiting for us in the shade of the café, hitting on the gift shop girls like nobody's business. Back in the car we hit the road north for Trang Bang. We pass through what's left of the Iron Triangle—the Ho Bo and Bo Loi Woods. We stop in Trang Bang for the hell of it and pop into a sidewalk billiard parlor where a bunch of young soldiers are playing snooker. During the war, when we were laagered at some forward support base in these parts, a couple tracks from recon would drive into this town and wait for the Cu Chi–Tay Ninh convoy. Then we would dismount and gather with our rifles and shotguns and whatnot at the café, drinking Ba Muoi Ba by the liter; generally taking a break in place, as the saying goes, under the abundant shade of the promenade plane trees, waiting for the morning convoy to come blowing through town. Trang Bang's square was one of the few places that seemed like a true respite from the war happening all around us otherwise, but now, today, it is nowhere to be seen.

We buy a round of beer and listen to the talk. I ask what happened to the fountain square. The players and loafers tell me that the fountain and square were done to death "with artillery" during the last days of the war as the VC and NVA worked their way south to Saigon. What a grind that must have been.

Soon enough we're back on the road and hang a right at Junction #22 and head north toward good old Tay Ninh. It may well have been the time of year, but the province of Tay Ninh looks absolutely parched. The rubber trees are, simply, gone. There is a weariness to everyone's step, but not because of the sharp oppression of the midday heat; it is as if what little wealth comes to Vietnam never reaches this far, and *that* you can see on the faces. It was almost painful to watch. Tay Ninh is the only province I visited where you saw oxcarts of the old kind; head-high wooden wheels with metal rims hauled with the blunt, dumb doggedness of oxen.

As we drive up Highway #22, a lightness comes into the car, a buoyancy, half-festive impatience for all the best reasons; downright cheer. I simply cannot keep my eye off the horizon, waiting for the first glimpse of the Black Virgin Mountain, La Montagne de la dame noire, the Nui Ba Den. Behind us, all the rest of the trip at our backs, we had been jolted by the exotic, the downright odd—all this the common experience of any interested traveler.

But the sense of anticipation, now, is of a different kind. Nui Ba Den is familiar to me, as "up-country" is not, and re-called with considerable warmth. Take it all around, I am in-trigued by my curiosity and much interested in what my response will be (as anyone ought); as if I could stand aside and watch myself from a distance—what Samuel Clemens called standing perpendicular to yourself. How *would* I respond, and am I old enough to distinguish the gut's reaction from senti-mentality. Simply put, I couldn't wait to see it.

Then. There it is, the lumpiest little bump, a dark and smoky silhouette yonder.

Seeing the Nui Ba Den teases up an intense anticipation that does not have anything to do with the war, and I am not the only American who served in that neck of the woods with a strong sense of that; but I am getting ahead of myself.

*   *   *

Soon enough, we cruise into the southern neighborhoods of Tay Ninh, and we pull up alongside the compound wall of the Cao Dai Temple, the holy see of Cao Daism.

The temple is set in a compound that is the better part of a mile square. There's a road junction near the main gate; the road east takes you out to the mountain. When my platoon worked convoy escort, we would park around the junction, waiting to escort the trucks from Tay Ninh base camp to Dau Tieng, our own base camp south and east about twenty miles yonder. I recall standing on the deck of my track (drawn by the atonal hum of many voices at noon prayers), looking over the compound wall past the seminary dormitories, the broad grounds, and to the tall block-long, bizarrely colorful temple. It has the look of a gingerbread house. Many a time I had wanted to walk inside and take a look at the temple close up, but Americans were not allowed inside the gate; we weren't allowed to enter the Vietnamese Catholic churches, either. During the war, it seemed to us that soon after a large firefight, several Cao Dai funerals would take place, and we would drive past the procession as it made its way to the cemetery—the body carried on a caisson in the form of a brightly painted boat and accompanied by brightly dressed musicians and mourners, everyone wearing white headbands. The temple is one of the few places I actually *wanted* to see on the trip.

We arrive just before noon, just in time for the daily noon service; the only Westerners anywhere near the place. One of the deacons, dressed in a long white tunic and cap, invites us to mount the stairs at the back to the mezzanine. Larry and I share the space with a five-piece orchestra and a children's choir, which plays and sings the strangest, droning music I have heard in a good long time.* The Cao Dai religion began in the 1920s, a strange and beautiful amalgam of the political ethics of Confucianism, the present-moment gentleness of Buddhism, the natural harmony of Taoism, and the morals of Judaism, Christianity, and Islam as well as Geniism (an indigenous religion of spirit worship similar to the Shinto of Japan). Among the pantheon of Cao Dai saints are Sun Yat-sen (leader of the Chinese Revolution of 1911), Victor Hugo (of all people), and Trang Trinh (a fifteenth-century Vietnamese poet). There are a number of others, including Jesus of Nazareth, Joan of Arc, Descartes, Shakespeare, Louis Pasteur, and Vladimir Lenin.

Above us, the vaulted ceiling is painted sky blue with bits of mirrored glass set in the plaster. Looking across that vast space toward the altar, you have the clear sense of being inside a meditation, that spiritual atmosphere common to the large interiors of European medieval cathedrals, the great mosques of the great cities of the Middle East, the great halls of the Forbidden City in Beijing, and the like. But that texture of great austerity in the cathedrals, the calm and intricately plain aesthetics of the mosques, simply cannot hold a candle to the riot of color and image intrinsic to the holy see of Cao Dai in Tay Ninh. Despite the blizzard of distractions, a strong spiritual sense pervades the temple, and you get a sense that *here*, in this place, beseech-

---

* *This years before busloads of day-tripping tourists would motor up from Saigon, hang around for the noon service, and then motor straight back to Saigon in time for cocktails. Nowadays the noontime service has been transformed into a performance—like the hotel-lobby hula dances of Hawaii or the museum-sponsored potlatches of the Pacific Northwest.*

ing Victor Hugo is in the same category as heartfelt, beseeching prayers to Saint Jude or even Saint Expedite. (How Victor Hugo would respond to prayers is a story for another time.)

While we lean on the railing, listening to the choir and orchestra, the congregation sits cross-legged on the floor, men on one side and women on the other, chanting prayers. That strong midday light of Cochin China, and the heat, comes hard through the tall windows and wide doorways, making the gowns and robes and tile floor glow. The droning nasal chants and music and singing fill the place—several hundred voices. Flights of swallows swoop and dip from one end of the temple to the other, circling the stout pink pillars (spirally wrapped with green dragons flashing long red tongues and bigger around than you can reach) and hawking like nightjars. What to make of this? What a place! A friend of mine once said that it was like being on the inside of a banana split.

Back in the car, we head east for Nui Ba Den.

Not far down the road is the Tay Ninh hospital, where, during that first writers' trip in 1990, we were conducted on a tour of the compound. The operating rooms were equipped with only gurneys and goosenecked lamps, there was a distinct shortage of surgical gloves, reusing needles was routine, and the doctors stropped their own surgical tools by hand. At the back of the main building we were shown a large storage room that smelled powerfully of formaldehyde and vinegar, raw wood and dust. The doctor flipped the switch to show us long shelves, stacked ceiling high with row after row of fetuses in gallon pickle jars—stillbirths, spontaneous abortions, and miscarriages. The result, we

were told, of Agent Orange spraying by Operation Ranch Hand.*
The birth defects displayed in that room, out of all normal pro-
portion to the population, were widely understood to be the result of the American wartime program of herbicide spraying. A
large head, floating in pickle, squeezed against the jar (eyeballs
like marbles); bodies pinched together, pale and raw, as if
pressed together by large hands; lumpy coils of bowel curled
around a torso; mouths agape as if in utter disbelief. When I was
a soldier I saw a thing or two, but I am also a father, and the contents of those bottles were extremely difficult to look at, much
less dwell upon. Some years ago I had the opportunity to see a
Ranch Hand plot map, and this area of Vietnam around Tay Ninh
was one large black splotch of ink. I don't recall that we were ever
told of Operation Ranch Hand, though it was hardly a military
secret, and more than once my platoon drove through woods that
looked freshly dusted, where everything was covered with a kind
of dirty-looking talc. Ugh.

That was the first writers' trip. This time Ferguson breezes
by the hospital without so much as a tip of the hat.

About halfway between the Cao Dai Temple and Nui Ba
Den, Ferguson stops in a roadside village and insists we buy a
couple handfuls of joss sticks. He does not explain why.

Then he pulls up the road apiece and stops by a broad and
deep, well-kempt grove of plane trees and banyan and the odd
rubber tree. It is a cemetery, surrounded by a low compound wall,
where six of the ten thousand local guerrillas killed during the

---

* *From 1961 to 1971, Operation Ranch Hand sprayed 19 million gallons of herbicide to
defoliate the countryside and downright kill the woods and deny the VC cover. Of that, 11
million gallons were Agent Orange, which contains dioxin, said to be the most potent
poison ever developed. Dioxin still, you may be sure, deeply saturates the water table.*

American War lie buried; cremated and interred in small graves with the names and dates and nothing else; much the same as you would see in military cemeteries of the Civil War. Cemeteries the like of which we have passed, casually and mostly without comment, since we left Hanoi; the grounds here swept clean every morning by a couple little old ladies swinging long brooms of rough thatch, scythe-fashion. Ferguson all but shoos us out of the car. We get out and walk up to the monument altar. We pause and light our handfuls of joss sticks. They flame up like torches. After a moment, we douse the flames with a sharp flick of the wrist, and the joss smokes right up with that thick and heavy, temple incense smell. We plant them, richly smoking, among the other bunches of joss in the pot of sand; we take a moment, there, rubbing the smoke on our faces and in our hair, trying to imagine peace and quiet for six thousand souls.

Military cemeteries are places apart. There is a poignant atmosphere of the deepest sadness about them, an undeniable texture of grief. Gettysburg or Flanders or Normandy, and all those other such places around the world (take your pick); rows of individual graves or one large, long common trench. It is so, I suppose, because everyone buried there was—until the moment of his death—young and healthy, even robust. Simply put, the war dead of military cemeteries did not just die; they were put to death.

North of here, in the middle of the country, there seem to be many; many. But that is where Vietnam is the narrowest; from Vinh to Hue, Laos is only forty miles from the South China Sea. And there are both Vietnamese and French; the Legionnaires buried near to Highway #1 called in those parts the *rue sans joie* (the street without joy, which for the French it certainly was; Bernard Fall used that phrase for the title of his book about the

French war against the Vietnamese). And north and west of Hue City near Highway #9 out toward Khe Sanh is the Trong Son Cemetery, where ten thousand Vietnamese killed along the Ho Chi Minh Trail are buried; think of Arlington National Cemetery on the Potomac River across the way from Washington, D.C. (built deliberately on the grounds of Robert E. Lee's former plantation mansion so that no one could ever live in the house again).

As Larry and I walk through the thin pastel shade of the Tay Ninh Cemetery and among the diminutive headstones, we come upon an astonishing irony. We have been accompanied by a handful of kids from the nearby village; this has happened wherever we have traveled. And as we walk, almost lollygagging with enlarged melancholy, one of the kids kicks up, just there, some old, large-caliber brass ammunition casings and offers them to us.

Souvenirs of the war, he supposes.

One brass cartridge is obviously a 40mm, about the size of one of the youngster's forearms; the others are .50-caliber cartridges. How grisly to know that there had been a firefight right here, we suppose, in the midst of what is now a cemetery; perhaps there are a number of Vietnamese buried here, where they were killed.

The brass. A .50-caliber machine gun is a loud and heavy weapon, and has an effective range of a solid mile. I stood behind a fifty for the last five months of my tour, and when you fired it, always good long bursts up and down, up and down, rather than sweeping from side to side, your arms and face positively shivered, and the dust shook out of your hair; nearly the same sensation as when you stand near very large bass speakers, listening to extremely loud music. The 40mm, a high-explosive round, came from what we called a "Duster"; a

Korean War–vintage M-42 tank body with a large armored tub mounted in the turret ring to accommodate two 40mm antiaircraft guns—the "pom-pom" guns you see in the World War Two documentaries trying to shoot down a Japanese kamikaze before it whacks the *Yorktown* (the gunners feeding ammunition in four-round clips). The guns were called Dusters because of the great clouds of dust kicked up as they fired, such was the force of the muzzle flashes; as much dust, I expect, as a Huey kicked up while setting down in dry season. I don't think that Dusters were *ever* used as antiaircraft guns. They made a sharp, pounding, godawful noise and could mince someone's house flat or transform a wood line into splintered kindling in nothing flat. In my hand, the brass feels dry, gritty, and old. I have not touched a brass cartridge, live or spent, since the war.

No thanks, kid, I have all the souvenirs a body could want.

Back in the car, we take off; not even Ferguson can get out of there fast enough. Ahead of us, Nui Ba Den fills all the windshield. We pass the road that during the war took you around to the rock-crusher, where the Army engineers "made" road gravel out of mountain boulders, and you heard that grinding a long way off.

Soon enough we pull into the parking lot. Larry and I decide to save lunch for later. We must see the mountain and the Ba Den Temple first. Ferguson stays with the car, hanging out in the parking lot. We have the vague idea that the temple is not far, so we walk up the road, through a hamlet of hooches, and cross a narrow footbridge.

There are not many visitors, but not few either, and, we notice right away, we are the only Westerners.

We follow the wide footpath, now rising to the low slopes of the mountain, and arrive at the stairs. This is closer to the mountain than I have ever been. There is one of those shin-high, red-and-white mileposts to the side. It reads:

Ba Den

1 km.

Larry and I look at each other, no sweat.

The high and wide, meandering steps are constructed of large, oblong stone and laid side by side sort of helter-skelter-like. Each step is a good reach, one above the other, and the whole affair is steep (as much as 45 degrees in some places). There is steady traffic, coming and going. Even little old ladies, descending, drop nimbly among the steps and look more weathered than bothered. How hard can this be, eh? What could be simpler, just start climbing. For the record we count the steps, but soon the stone-step path is a jumble and we lose count; steep steps, almost flat places, chunks of rock obviously too large to move (the steps go "around" them).

Soon enough we are pouring sweat for the second time to-day. My legs still ache from duckwalking the tunnels. We pass first one open-air café and then another, and another. We give up counting the steps; it's a klick, let's let it go at that. Finally, we are so godawful *hot*, our breath so short, our legs so cramped that we *have* to stop. The next café we come to we just turn off the steps and walk right in and take seats in the breeze. The woman who runs the place and her daughter have seen this all before. Without so much as a how-do-you-do, they dip into their Styrofoam cooler and fetch four one-liter bottles of water and two soaking-wet ice-cold wash towels. The woman drapes these

around our necks—ah!—and cracks open the first two water bottles for us; thank you, *ma'am*. These we drink right down; we also
drink the second bottles and ask for a third round. Larry and I
dawdle for a quarter of an hour, cooling down and watering up,
and then start out again. Twice more we stop for cold, sopping-
wet towels around the neck and ice-cold bottled water. Our pace
takes on the slow-motion doggedness of patient determination.
This, my man, is going to be one hell of a climb, and, of course,
will only get worse. (I don't believe I will ever get what some people in this world *see* in backpacking or camping—haul a big
bunch of stuff on your back into the woods and sleep on the
ground with the bugs; they must be out of their minds.)

About halfway up to the temple of Ba Den, we stop under the
most welcome shade of a red-blooming flame tree to catch our
breath for the umpteenth time. We look east and south and west,
a broad panorama of countryside. For all our climbing, this is the
first moment I stand still, turn my head, and *look* out across the
countryside toward the horizon. From the shade, Larry and I can
see about 220 degrees of arc on the horizon. (Imagine the hands
of a clock; you looking at the six, the hour hand pointing straight
at three and the minute hand just past ten.) I grew up in the Midwest, and don't get to see the world like this very often.

The whole sweep of countryside is parched, and as flat as
the back of your hand; so flat, as we say in the Midwest, that you
can see the company coming three days off.

Here before me lies my war-year of soul-deadening dread
(the whole fucking thing, as if compressed in dream) as I have
never seen it before, and I am standing in the middle of it, like
a bull's-eye.

Every place I ever camped; laagered, humped an ambush;

got drunk, got stoned, got laid. Every place we ever busted jungle; every firefight gone bad; every monsoon downpour we sat through, stoic and grim, the rain blowing *down* at us; every little bitty village we ever took off after; every month-old, flyblown corpse we ever found. The Cao Dai Temple and Tay Ninh and the cemetery is there (point west); there (point south), the Michelin Rubber plantation and Dau Tieng—the area around my platoon tents now a cashew farm.

Coming straight at us below is the road from the village of Soui Dau; just there to the left across the creek is the brickyard, French fort, and Firebase Grant. The road isn't any wider than a country lane and resembles a California dry-season firebreak. At the foot of the mountain, just here, it abruptly turns east and continues around the mountain, then straight north fifteen miles or so (as the crow flies) to Soui Cut.

Soui Cut, not five miles from Cambodia and not much more than a wide place in the road, where on the night of January 1, 1968 (by remarkable coincidence), filmmaker Oliver Stone's battalion and my battalion with two batteries of 105 howitzers were attacked by the 272nd NVA Regiment, and we killed five hundred guys in one night—and, trust me, it took all night. It has to be the longest night of my life, and without a doubt the worst; always called by us The Great Truce-Day Body Count.

My battalion left Dau Tieng the day after Christmas with a bulldozer and a Duster—those twin-mounted 40mm guns—and the minute we arrived at Soui Cut we knew something was up; the tail end of the Ho Chi Minh Trail was just north of us through the woods, after all. The battalion of straight-leg grunts, with Oliver Stone among them, took up positions on the east half of the perimeter, and my battalion of tracks took up positions on the west with the road south to Highway #13 in between. The night of the big battle the recon tracks with our thirty machine guns were parceled out among the line companies, and I moved in with a platoon of Bravo Company tracks facing north (not thirty paces from the wood line). Bravo Company, the hard-luck company of the battalion, was almost all black guys, and they did not look happy; but then no one was happy. We pulled up to the perimeter and parked; pulled up ammunition for the machine guns, our boxes of hand grenades, the bag of shotgun rounds; set out all our claymores; idled the engine for an hour to make sure the batteries were charged (for the radio), and called it a day. We used a nifty trick with the claymores. It was not unheard-of that some wiseacre would crawl up to the laager during the night and simply help himself to one or more. So we would booby-trap a couple of them. We'd take a frag, switch the four-and-a-half-second fuse with a second-and-a-half fuse from a smoke grenade, dig a cup-sized hole in the dirt with a bayonet, pull the grenade pin (careful to save it; you'll need it in the morning, cousin), set the frag on its side, place the claymore on top to hold it down, camouflage the frag with dirt and trash and such, and then leave it alone until the morning. Then if someone tried to grab one, well, that would be the end of that. If you actually had to blow the claymore, well, the frag would give it extra oomph.

By dark everything was quiet. The straight legs sent out an ambush; they gathered at the other end of the laager and left. Many three-man listening posts (LPs) moved into the woods, found cover, and set up; they would sit there all night, quiet as mice, and radio in anything that came their way. Around eleven the ambush popped and there was a brief, intense firefight—the racket coming through the woods from a good way off—and everything was quiet again. Not long after, one of the LPs out in front of us called in, whispering, and said he had movement— many, many bodies moving past him through the woods. Oh. We settled in behind our guns; me on the fifty, the gunners in back at their M-60s, and the driver with his shotgun, grenades, and the claymores. I don't remember how it began; all I remember is that around midnight we just started in. RPGs and mortars and sustained volleys of AK-47 fire. Claymores and grenades and fifties and M-16s and shotguns and .45s and artillery fire from right behind us, just there, and the Duster pounding away at the side of the road south of us. Low voices on the radio reverberated around the laager (like the public announcer's voice at the ballpark), voices out loud—both English and Vietnamese. We fired at movement and muzzle flashes, at anyone who came out of the woods, just there; simply said, we shot at them as they came toward us. We went through box after box after box of ammunition. We were told that the NVA had made it inside the perimeter over by the straight legs and that I should get ready to move, but we never did. There were F-4 Phantom air strikes from Tan Son Nhut; the colonel made the call and ten minutes later the planes arrived with bombs, napalm, and strafing runs for lagniappe. There were helicopter gunships and long-range heavy 8-inch guns and 175mm cannon. Somewhere in here we were told to take cover, because the artillery right behind us was going to fire a salvo or two of point-blank beehive flechette rounds— grapeshot in the form of thousands of 4p finishing nails.

The shit flew all night. I clearly recall being startled when the sun finally came up and we could finally look out down front; corpses everywhere, bare feet and flies, the wood line shot to pieces, and firefight trash everywhere. Everything covered with dust, antique-looking.

A patrol was sent out to look for the ambush and brought all eight of them back on stretchers, all dead. The Vietnamese must have washed over them like a seiche rolls over a beach.

The bulldozer dug a shoulder-high ditch, and we spent the morning burying, for sanitation's sake, the Vietnamese dead; bodies and parts of bodies. We did it like you'd make lasagna— a layer of bodies and body parts, a generously thick broadcast of quicklime (hastily choppered-in earlier that morning along with the reporters), another layer of bodies, and so on. When we were finally finished, the bulldozer filled in the ditch with dirt and firmed it down the way they do, with a couple passes back and forth; even so, the smell seeped up, gathered, and lingered among us, and simply did not go away. That was the night the Tet Offensive of 1968 began, for us, in earnest. I am told we made the front page of the *New York Times*.

The battle scene served as the model for the climactic scene in Stone's film *Platoon*, one of the first Vietnam War films to convey a true texture of the war as a place. In my first novel, the battle was spoken of in the chapter called "By the Rule."

So, Larry and I stand under our flame tree, and I'm saying to myself: What on earth was I thinking to come *here?*

To answer that, first of all let's get one thing straight. Coming back is not the wistful nostalgia of some geezer of an "old soldier" hankering for the "old days." Why would anyone want to travel again back through the black years that have reverberated ever after to *that* death-green place? The unmistakable, perpetual itch of something crawling on your body, through your hair, burrowing into skin; humping a seventy or eighty or hundred-pound pack crammed with absolutely everything you will need right now, tonight, first thing in the morning, and not much else; the tangy, salty sweat clinging to every shaving cut, every scratched-to-death mosquito-tick-lice-crab bite (each and every sweat pimple and sore-that-will-not-heal); the malaria pill induced diarrhea that will not go away (your anus a raw infection all by itself); the endless pick-and-shovel shit-work of a kind that is hard to duplicate (unless you're talking about big-cut logging or stoop-labor migrant work, pit-mining or offshore oil drilling). Never mind the bloody murder, *every-thing* was a grind. The smell of your own body that ripens for a week while your clothes all but rot off your back (ever been around someone who simply will not bathe; that dull and fetid, greasy smell?), then all but disappears for a week, and then reemerges as something just plain foul. The shit-scared panic of a firefight gone all wrong. The fuckups, the blunders, and picking up the aftermath trash. The gradual, palpable disap-pearance of ordinary, natural feeling (*all* the organs of physical perception). Every human vitality is taken from you as if you'd been skinned; yanked out like you pull nails with a claw ham-mer; boiled off, the same as you render a carcass at hog killing.

As well, you must understand, I don't come back to "heal," but I can only speak of it this way. In Melville's *Moby-Dick*, toward the end, as the story is coming down to the button, Captain Ahab has the blacksmith fashion a handful of horseshoe nails and steel razors into harpoon tips, which Ahab will daub with blood coldly donated by the crew; it is with these "irons" (spicy with the mal-ice of vengeance) that Ahab intends to kill the hated white whale.

As the smithy bangs out the new tips red-hot at the forge, Ahab asks him with the true innocence of drama whether the man is afraid of getting burned in that blast of sparks; and the smith replies, quite bluntly (speaking for many a man who has made his bread by the swing of his arm), that, no, he is not afraid of getting nipped, because, Cap'n, you cannot scorch a scar.

If I wanted to soak myself in the war, to drape it around me, and disappear into that wallow of grief, all I would have to do is visit the Vietnam Veterans Memorial in Washington. Panel 46 East, that wing of the memorial pointed directly at the Washington Monument; *there* are the score of names I know. I have heard it said that the memorial is the focused expression of national grief that surrounds our war in Vietnam, and is the most visited place in Washington; for more than twenty years it's been famous. And when you go to that place, there must be something about being a veteran that tips it off. Maybe it's our beards and paunches, the beads and bikes, the three-piece suits and union shirts; apparently there is something that comes into the eye, but I'm not talking about the crackle of wrinkles or the blinking moistness that too soon becomes tears. I'm talking that numb, long drawn-down iron look; that flat iron color of the face; a roundness to the shoulders and that numb iron pace of step that is a personification of flat slabs of dark granite. As it happens, there is a Vietnamese legend of the warrior hero of ages past who was given an iron horse, a great iron whip, and a suit of iron clothes. And the very instant he threw a leg over the horse's back, *snap!*, the horse came alive and began breathing fire, and the man's heart turned to iron. Of all the accoutrements in a soldier's kit, an iron heart serves best, just then. The work of soldiers is, finally, too mean for anything else. Iron was once thought to be magic, and smithies at the forge, magicians; something so hard that did not give to the touch, made with fire, fashioned with sparks, and tempered with a hiss. Magician's work, indeed.

So, the iron in the eyes is not easily masked, even behind metallic tinted sunglasses.

At the memorial there is an intentness on the names; looking for that handful of guys you haven't laid eyes on since the day they were killed, or last saw the day you left; and your vivid-as-a-dream memory of them: the incompetent medic who took canned fruit and a real can opener on ambush, but could be absolutely counted on to come get you no matter where you were or how bad you were hurt; the built-like-a-fireplug, high-school-wrestler motorcycle punk, and the platoon character; the guy who was unremarkable in everything, but who burnt to death at Soui Cut when his track took an RPG and we didn't find enough of his body to fill a sandbag; the petty-minded fathead who got squeezed in half when his track drove over a footlocker filled with explosives buried in the road, and the track flipped over backward on him; and the like. Not dead heroes, mind you, just dead, and for reasons so small that it makes your heart sore and your eyes ache at the thought of it.

But there is a larger irony here that the casual visitor, perhaps, does not immediately appreciate. Across the street and past the Lincoln Memorial, over the Potomac River, and up the slope next to Arlington National Cemetery is, probably, the most remarkable image of a war memorial in all of American history; the United States Marine Corps Memorial, more commonly referred to as the Iwo Jima Bronze.

Within sight of each other are these two extraordinary, virtually opposite images of war; each evocative, and both all but overwhelming. The Iwo Jima Bronze is *all* work. Those guys spent how many weeks aboard some Navy transport; everyone sick as dogs, and no grabbing a smoke on deck after dark. They arrived, strapped it on the next morning, climbed down cargo nets thrown over the side, and loaded into landing craft that will not be still. At the general's word, the Higgins boats made for the beach (hauling ass), struck shallow water, the front ramp popped open, everyone piled off and ran up the beach for their very lives to confront 22,000 well-dug-in Japanese. It took 70,000 Marines thirty-six days to occupy an island eight miles square. There were something like 190 Marines killed per day, and only a handful of Japanese survived—death before dishonor. Iwo Jima was a dumb lump of sulfurous volcanic scrap; and we can imagine that the smell only got worse as the battle of the island dragged on. Then one afternoon, a bunch of guys scratched their way to the top of Suribachi and hoisted a flag.

Stand beneath the Iwo Jima Bronze, looking *up*, and you understand very quickly that all the energy of effort is coming *at* you. The sheer physical gesture of the pose (an intimation of war as work)—that one guy's hand and fingers reaching, not quite touching. And one more thing: you cannot touch the Iwo Jima Bronze.

But stand at the Vietnam Veterans Memorial, and the energy comes *from* you, and you *must* touch it. The invitation to touch is more than an intimacy with familiar names, brothers of the blood; everyone is compelled. The granite is just there, the names just there; it is irresistible. When the memorial was first built, it was heavily, even bitterly, opposed because it was "below ground," because it was "a black gash" (some called it), and was designed by "an oriental"—Maya Lin—though this criticism was kept extremely quiet. But I suspect that the real reason for the opposition was an edgy and skittish, sentimental reluctance to involve oneself physically with the memorial; much less personally or spiritually. No, no, the critics seemed to say, let's not have any of that; let's have something grandly martial and inaccessible and pointless, something we can march around and salute.

You walk into the Vietnam Veterans Memorial, passing those slabs of black granite deeply incised with tens and hundreds and thousands of names—first one row of names, then five rows, then ten, then fifteen, and you read them, but then the names come thick and fast, and become finally a blur.

To stand at the crease in the middle of the memorial, the list of names higher than you can reach, where the very first and the very last men killed are virtually side by side. Well, as I said, all the energy comes from you; and this because of the peculiar phenomenon of the reflections discovered only after the memorial was built. But even this matter of the reflections has a curious contrariness to it. The granite, from Bangalore, India, is as black as the ace of spades, polished to a high luster, and during the day when the sun is above and behind you, you see very clearly your reflection behind the names. That is spooky enough, but what really cured *me* of going to the memorial was being there at night. At dusk and dark, the tourists depart, and the halogen footlights under thick glass come on, the kind you see in a municipal swimming pool. These lights shine up in scalloped shapes, like footlights across a stage. And the effect is that your reflection is in *front* of the names. Very spooky, indeed.

Native Americans of my acquaintance tell me that *that*

place has been transformed into what is called a "power spot"; that the place now exudes the power of spirit of 58,000 souls. Well, I'm a city kid, so I don't know about that; all I know is that strange things happen when you invoke the names of the dead in such overwhelming numbers. I understand that each Veterans Day all 58,000 (plus) names are read aloud, and that the recitation takes several days.

So, you stand there, wallowing in that irresistible welling-up of irremediable grief. You stare at the granite, those several names familiar to you deeply and sharply incised (and your clear, dark reflection), and you spread your hands out to touch the names (*leaning* on them), and the tips of your fingers grasp the raw granite texture of the very letters, and the whole fucking war comes pouring out of you as if you're drunk sick—you cannot fucking help it; perhaps *this* is what Chief Seattle was driving at when he spoke of a sympathetic touch. You stand there a good long time, squeezing your hands against the granite. But then you look again, and, as poet Yusef Komunyakaa has said, it is only a woman brushing a boy's hair.

Meanwhile, the tourists are gathered six deep around you, watching with curiously disengaged, theatrical appreciation and a kind of awe, waiting for you to launch into one of those crying jags that veterans at the memorial are famous for. Sad to say, the memorial is also quietly famous for suicides, but those happen at night (a story for another time).

It is an old, old human wisdom that goes well back into the long millennia of the Great Hunt that when the slaughterhouse offal is sluiced away and the meat is dressed and hung, but no prayers are shouted aloft to accompany the spirit of the thing killed, then the grief of your murders is not assuaged; your heart of iron turns black, dead places literally appear on your heart, and the patina of grief poisons all the senses.

And, trust me, a recitation of the names—even so grand a thing as engravings in granite—is not enough.

NO, I don't have to travel halfway around the world to "see" the war, but to be rid of it.

Rather, it is Nui Ba Den, and *not* the war, that draws me here. To begin with, the Black Virgin Mountain, La Montagne de la dame noire, Nui Ba Den, is a place, a literal spot on the map, 996 meters high—almost a solid klick. It was as if you placed Mount McKinley in the middle of Kansas. I have talked with many guys who served in that neck of the woods, and almost to a man they say that their image of Nui Ba Den does not have anything to do with the war; yeah, they will say, I remember the Black Virgin Mountain—and a light comes into their eyes. It is almost literally true that we could see the mountain from everywhere we operated, everywhere we camped, everywhere we sat ambush, every firebase—all you'd have to do is look up and there was this mountain, this presence, this silhouette (the color of antique silver at night in the moonlight).

Now, I have traveled Vietnam from one end to the other, and when I'm asked where I served during the war I mention Cu Chi, Dau Tieng and Tay Ninh, and Nui Ba Den. Well, everyone has heard of Cu Chi because of the tunnels; Dau Tieng means the Michelin Rubber plantation; everyone knows of Tay Ninh because of the Cao Dai Temple; but absolutely everyone, north and south, man or woman, young or old, scholar or buffalo boy knows of Nui Ba Den because of the story. And, as I take it, the story is *very* Vietnamese.

Actually, I have heard several stories.

One story goes that when Ba Den's soldier-husband was killed in war, she so grieved his loss that she wanted to join him, so she climbed the mountain to be as close to heaven as she could manage and then committed suicide. Another story tells of Ba Den and her soldier-husband living at the foot of the mountain, and while he was away fighting, she was captured and raped, and afterward died of shame. The third story relates that though she was unknown in those parts but renowned for her spiritual simplicity, Ba Den would climb to the summit to meditate and pray, and when she died the mountain was so named to honor her anonymous and humble piety. Another story asserts that she was a historical person who lived in the later part of the eighteenth century. Ba Den was born into a family of poor farmers, of which in that neck of the woods there were plenty, and grew into a beautiful young woman who had a sweet nature and long black hair. She fell in love with a young soldier, but also caught the eye of the son of a rich merchant from Trang Bang. The two young men got into an argument and had it out, and of course the soldier kicked the other guy's ass. The soldier was then called away to war, the parting was, of course, bittersweet, and Ba Den promised to wait for him. The merchant's son, meanwhile, contrived to have her for himself and had her kidnapped, but she escaped up the mountain and, rather than submit, threw herself off one of the many stone cliffs and committed suicide. Her body was found by Buddhist monks, she was properly buried, and a temple was built in her memory. Some years later she appeared in the dream of a general who was about to begin a battle in those parts, told him how to win the battle, which he did, and afterward he had a bronze statue of her placed there. The mountain, theretofore known as Nui Moi (One Mountain), was renamed Nui Ba Den.

But the story I favor is this: when she was a young woman, Ba Den was to marry a soldier, but on her wedding day her husband-to-be was called away to war (the story does not say which one) and never returned. Yet Ba Den waited for him, cried so hard in her grief and longing that her family thought she would lose her eyesight, and, as the legend has it, when she died, her spirit *became* the mountain.

Larry and I finally, *finally* approach the top. The last hundred steps or so take forever, but when we come to that last step and arrive at the stone pavement of the temple grounds, well, our relief is audible to everyone within earshot.

The Ba Den Temple is not a static artifact of Vietnamese history, and *hardly* a tourist site, but rather a functioning Buddhist pagoda and monastery, and a place of serious pilgrimage and worship; the story of her faithfulness strikes a deep chord in the Vietnamese; *The Tale of Kieu* by Nguyen Du of the early nineteenth century is, after all, regarded as the classic epic poem of all Vietnamese literature, and the story is much the same; there are many such lovers' stories in traditional, popular, and folk literature. During the Tet Lunar New Year as many as fifty thousand people will visit the place, to honor the memory of Ba Den and pray for her blessings of patience and fortitude.

First of all, you have to understand that for someone like me who grew up on the prairie, mountains hold the fascination of surprise. In all weathers, day or night, moonlight or no, clear as glass or deep monsoon overcast, Nui Ba Den was, simply and forthrightly, well, *there*; large; beautiful to look at.

It goes like this. When I was a soldier in these parts, I always tried to take the last night guard, sitting in the main hatch of my track behind the fifty, from two or three until breakfast. I would watch processions of Viet Cong climb this way and that around the mountain, each man carrying a tiny perfume-bottle lamp, each light not much more, I expect, than the light of a birthday candle; the mountain seemed to shimmer at such times. Or take those nights one side or the other of the New Moon, when there was so little light the heavens seemed utterly vast, with only those many little lights of stars; the nights when you swore you couldn't see your hand in front of your face; but there the mountain would be, a faint and solid silhouette, as fuzzy as an apparition. And then there were those mad, murderous nights of fighting. The killing would cease only when the sun rose, the smoke cleared, and the dew burned off; and in that aftermath of berserk exhaustion, you looked down in front of you and there was nothing but meat and the wood line looking like ruined drapes. And then you looked out across the way, and *there*, rising sharply above everything, was Nui Ba Den.

But sitting up in that last long hour before dawn, the mountain—there—cut a clean silhouette in the scrub of dirty dark. Soon it was a blunt, Brunswick green. Then a peculiar grayish green as the light gathered above us, the mountain rough-textured with the rubble of mold-stained boulders and thick stands of weathered timber. Then a transparent seedling green; the sky not quite blue. Then—*boom*—the sun would rise like a big fat fire, the world once again all color. And there it would be, Nui Ba Den, vivid and entire. The greenest of all green.

For a Midwest city kid, Nui Ba Den was an enchanted image, but the mountain touched all of us with something we didn't even know we needed, and distinctly did not have anything to do with the war, but which soaked in more deeply and surely (touch your heart) than anything that happened back then.

Larry and I buy more joss sticks from a handy vendor, walk into the temple, and join a close crowd of worshipers come to pray in front of the seated image of Ba Den, complete with that enigmatic Buddha smile of bright porcelain calm. To each side of the statue are two tall, long-legged bronze cranes standing on the backs of turtles; images common to all Vietnamese Buddhist temples. The crane and the tortoise are emblematic of an *old* Vietnamese story of the faith of friendship; about everyone getting what they need. And the turtle, of course, is the mythical beast that brought knowledge and wisdom into the world. In front of the Ba Den's Buddha statue is a bronze bowl, about the size of a wedding punch-bowl, filled with sand. Here the worshipers plant their handfuls of joss—always an odd number (even numbers are bad luck). The sharply pungent smoke hangs in the air and fills the room. Strings of gray ash litter the altar and the stones underfoot. To one side is a flat platter for offerings, given in Vietnamese dong notes (pennies, nickels, and dimes) and plates of fruit and sticky-rice cakes (garnished with an almond and wrapped in elegantly bound palm leaves). A monk dressed in a workaday brown robe stands next to a bowl-shaped temple bell, holding a wooden clapper. Larry and I stand head and shoulders above everyone else, and the guy eyes us particularly. Every time someone steps out of the crowd and reaches up to place their smoking sticks of joss, the monk strikes the bell, and that deep and solid *boom* fills the room (despite the close softness of the crowd). A long moment I stand there, thinking that it all comes down to this. The blessings of your life sought; discovered; stumbled upon; *given* to you, as if pushed into your hand. Beginning with the simple fact of your life; any soldier will tell you that. You haven't blown your brains out; you haven't boozed yourself to death; Agent Orange hasn't incinerated your liver with cancer; you're not in the Lifer's Club at Joliet. I stand there so long, in fact, it becomes a meditation. And when I step forward, like the rest, to plant my several sticks of joss, the monk nods, signifying with a large smile and a long wink, and whacks the bell a swift and righteous lick. That hardy boom reverberates up into the smoky rafters.

Outside, shortly afterward, Larry and I stand at the low compound wall around the edge of the thick stone patio; the temple, the constant murmur and hubbub of the crowd at our backs. We look out across the wide, hazy panorama of the hot Tay Ninh countryside, taking it all in; talking about the whole trip, and other things besides. While we talk, the clear sense rises in me, bursting on me in a rush of honest revelation; and how odd a sensation. I'm home, I say to myself; I have arrived home; *this* place is home. Larry and I look at each other, and I am almost embarrassed at the discovered clarity. Home. Well.

Larry, I say, what do you make of that?

About the Author

Larry Heinemann is the author of three novels: *Close Quarters* (1977); *Paco's Story* (1987), winner of the National Book Award; and *Cooler by the Lake* (1992). He lives in his native city of Chicago.

AFTERNOON OF THE ELVES

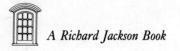

A Richard Jackson Book

ALSO BY THE AUTHOR

The Dancing Cats of Applesap

Sirens and Spies

The Great Dimpole Oak

AFTERNOON
of the ELVES

ƙ ƙ ƙ ƙ ƙ ƙ

Janet Taylor Lisle

Orchard Books · New York

A division of Franklin Watts, Inc.

RETA E. KING LIBRARY
CHADRON STATE COLLEGE
CHADRON, NE 69337

Text copyright © 1989 by Janet Taylor Lisle
All rights reserved. No part of this book may be reproduced or
transmitted in any form or by any means, electronic or
mechanical, including photocopying, recording or by any
information storage or retrieval system, without permission in
writing from the Publisher.

Orchard Books, a division of Franklin Watts, Inc.
387 Park Avenue South
New York, NY 10016

Manufactured in the United States of America
Book design by Mina Greenstein
The text of this book is set in 12 pt. Janson
10 9 8 7 6 5 4 3 2 1

Library of Congress Cataloging-in-Publication Data
Lisle, Janet Taylor.
Afternoon of the elves / Janet Taylor Lisle. p. cm.
Summary: As Hillary works in the miniature village, allegedly
built by elves, in Sara-Kate's backyard, she becomes more and
more curious about Sara-Kate's real life inside her big, gloomy
house with her mysterious, silent mother.
ISBN 0-531-05837-9. ISBN 0-531-08437-X (lib. bdg.)
I. Title.
PZ7.L6912Af 1989 [Fic]—dc19 88-35099 CIP AC

+
L689a

8 20

Jordania

5-14-90

For Elizabeth

a friend to elves

One ✗   The afternoon Hillary first saw the elf village, she couldn't believe her eyes.

"Are you sure it isn't mice?" she asked Sara-Kate, who stood beside her, thin and nervous. "The houses are small enough for mice."

"No, it isn't," Sara-Kate said. "Mice don't make villages in people's backyards."

Hillary got down on her hands and knees to look more closely. She counted the tiny houses. There were nine, each made of sticks bound delicately together with bits of string and wire.

"And there's a well," she whispered, "with a bucket that winds down on a string to pull the water out."

"Not a bucket. A bottlecap!" snorted Sara-Kate, twitching her long, shaggy hair away from her face. She was eleven, two years older than Hillary, and she had never spoken to the younger girl before. She had hardly looked at her before.

"Can I try drawing some water?" Hillary asked.

Sara-Kate said, "No."

The roofs of the houses were maple leaves attached to the sticks at jaunty angles. And because it was autumn, the leaves were lovely colors, orange-red, reddish-orange, deep yellow. Each house had a small yard in front neatly bordered with stones that appeared to have come from the driveway.

"They used the leaves dropping off those trees over there," Hillary said.

Sara-Kate shrugged. "Why not? The leaves make the houses pretty."

"How did they get these stones all the way over here?" Hillary asked.

"Elves are strong," Sara-Kate said. "And magic."

Hillary looked at her suspiciously then. It wasn't that she didn't believe so much as that she couldn't right away put Sara-Kate on the side of magic. There never had been one pretty thing about her. Nothing soft or mysterious. Her face was narrow and ended in a sharp chin, and her eyes were small and hard as bullets. They were such little eyes, and set so deeply in her head, that the impression she gave was of a gaunt, fierce bird, a rather untidy bird if one took her clothes into consideration. They hung on her frame, an assortment of ill-fitting, wrinkly garments. ("Doesn't she care how she looks?" a new girl at school had inquired just this fall, giving every child within earshot the chance to whirl around and shout, "No!")

Least magical of all, Sara-Kate Connolly wore boots that were exactly like the work boots worn by men in gas stations.

"Black and greasy," Hillary's friend Jane Webster said.

"She found them at the dump," Alison Mancini whispered.

2 ·

"No she didn't. Alison, that's terrible!"

Normally, fourth graders were too shy to risk comment on students in higher grades. But Sara-Kate had been held back in school that year. She was taking the fifth grade all over again, which made her a curiosity.

"Can you tell me where you found those amazing boots? I've just got to get some exactly like them," Jane said to her one day, wearing a look of such innocence that for a second nobody thought to laugh.

In the middle of Sara-Kate's backyard, Hillary recalled the sound of that laughter while she stared at Sara-Kate's boots. Then she glanced up at Sara-Kate's face.

"Why does it have to be elves? Why couldn't it be birds or chipmunks or some animal we've never heard of? Or maybe some person made these houses," Hillary said, a sly tone in her voice. She got off her knees and stood up beside the older girl. "We are the same height!" she announced in surprise.

They were almost the same except for Sara-Kate's thinness. Hillary was sturdily built and stood on wide feet.

"In fact, I'm even a little taller!" Hillary exclaimed, rising up a bit on her toes and looking down.

Sara-Kate stepped away from her quickly. She folded her arms across her chest and beamed her small, hard eyes straight into Hillary's wide ones.

"Look," she said. "I didn't have to invite you over

here today and I didn't have to show you this. I thought you might like to see an elf village for a change. If you don't believe it's elves, that's your problem. I *know* it's elves."

So, there they were: elves—a whole village of them living down in Sara-Kate's junky, overgrown backyard that was itself in back of Sara-Kate's broken-down house with the paint peeling off. Sara-Kate's yard was not the place Hillary would have picked to build a village if she were an elf. Where there weren't thistles and weeds there was mud, and in the mud, broken glass and wire and pieces of rope. There were old black tires and rusty parts of car engines and a washing machine turned over on its side. Carpets of poison ivy grew under the trees and among the bushes. Nobody ever played in Sara-Kate's backyard. But then, as Sara-Kate would have said, nobody had ever been invited to play in her backyard. Except Hillary, that is, on that first afternoon of the elves.

𝕜 "Sara-Kate Connolly thinks she's got elves," Hillary told her mother when she came home, rather late, from looking at the village. The yards of the two families backed up to each other, a source of irritation to Hillary's father, who believed that property should be kept up to standard. But who could he complain to? Sara-Kate's father did not live there anymore. ("He's away on a trip," Sara-Kate always said.) And Sara-Kate's mother didn't care about yards. She hardly ever went outside. She kept the shades of the house drawn down tight, even in summer.

"Elves?" Mrs. Lenox repeated.

"They're living in her backyard," Hillary said. "They have little houses and a well. I said it must be something else but Sara-Kate is sure it's elves. It couldn't be, could it?"

"I don't like you playing in that yard," Hillary's mother told her. "It's not a safe place for children. If you want to see Sara-Kate, invite her over here."

"Sara-Kate won't come over here. She never goes to other people's houses. And she never invites anyone to her house," Hillary added significantly. She tried to flick her hair over her shoulder the way Sara-Kate had done it that afternoon. But the sides were too short and refused to stay back.

"It seems that Sara-Kate is beginning to change her mind about invitations," Mrs. Lenox said then, with an unhappy bend in the corners of her mouth.

But how could Hillary invite Sara-Kate to play? And play with what? The elves were not in Hillary's backyard, which was neat and well-tended, with an apple tree to climb and a round garden filled with autumn flowers. Hillary's father had bought a stone birdbath at a garden shop and placed it on a small mound at the center of the garden. He'd planted ivy on the mound and trained it to grow up the birdbath's fluted stem. Birds came from all over the neighborhood to swim there, and even squirrels and chipmunks dashed through for a dip. The birdbath made the garden beautiful.

"Now it's a real garden," Hillary's father had said proudly, and, until that afternoon, Hillary had agreed. She had thought it was among the most perfect gardens on earth.

Sara-Kate's elves began to change things almost immediately, however. Not that Hillary really believed in them. No, she didn't. Why should she? Sara-Kate was not her friend. But, even without being believed, magic can begin to change things. It moves invisibly through the air, dissolving the usual ways of seeing, allowing new ways to creep in, secretly, quietly, like a stray cat sliding through bushes.

"Sara-Kate says elves don't like being out in the open," Hillary remarked that evening as she and her father strolled across their garden's well-mowed lawn. She found herself examining the birdbath with new, critical eyes.

"She says they need weeds and bushes to hide under, and bottlecaps and string lying around to make their wells."

Mr. Lenox didn't answer. He had bent over to fix a piece of ivy that had come free from the birdbath.

"And stones on their driveways," Hillary added, turning to gaze at her own driveway, which was tarred down smooth and flat.

She turned toward Sara-Kate's house next. Its dark form loomed behind the hedge at the bottom of the yard. Though evening had come, no light showed in any of the windows.

Now that Hillary thought about it, she could not remember ever seeing many lights down there. Gray and expressionless was how the house generally appeared. What could Sara-Kate and her mother be doing inside? Hillary wondered, and, for a moment, she had a rather grim vision of two shapes sitting motionless at a table in the dark.

Then she remembered the shades. Mrs. Connolly's shades must be drawn so tightly that not a ray of light could escape. Behind them, Sara-Kate was probably having dinner in the kitchen, or she was doing her homework.

"What happened at school today?" her mother would be asking her. Or, "Please don't talk with your mouth full!"

Hillary imagined Sara-Kate Connolly frowning after this remark. She felt sure that Sara-Kate was too old to be reminded of her manners. Too old and too tough. Not really the kind of person to have elves in her backyard, Hillary thought.

"I'm going inside!" Hillary's father's voice sounded from across the lawn. The rest of him was swallowed up by dark.

"Wait for me. Wait!" Hillary cried. She didn't want to be left behind. Night had fallen so quickly, like a great black curtain on a stage. In a minute she might have been quite frightened except that suddenly, through the garden, the twinkling lights of the fireflies burst forth. It was as if the little bugs had waited all day for this moment to leap out of hiding. Or had they been there all along, blinking steadily but invisibly in the daylight? Hillary paused and looked about.

"Hillary! Where are you?"

"Coming," she called, and turned to run in. A gust of wind slid across her cheek. Like lanterns in the grip of magic hands, the tiny lights flickered over the lawn.

TWO

Hillary dreamed about elves during the night. By morning it was clear that the magic of Sara-Kate's elves must be real, for while Hillary slept, it crept, mysterious and cat-like as ever, out of the Connollys' backyard, up the hill and through the half-opened window of Hillary's bedroom. There she woke beneath its spell shortly after dawn and immediately was seized by a mad desire to run down to Sara-Kate's yard in her nightgown.

But what would Sara-Kate have thought? And suppose the elves were not such early risers? Hillary imagined them surprised in their beds, leaping for cover as her giant bare feet thudded over the ground toward the fragile village. She made herself dress for school instead. She gathered her school books with unusual attention to orderliness and went downstairs to the kitchen. Hillary was determined to visit the elf village again soon, that very afternoon if Sara-Kate would have her. In the meantime, she ate a large breakfast of pancakes and milk, walked to school four blocks away, and spoke privately to Jane Webster and Alison Mancini about what she had seen in the Connollys' yard.

"Elves!" shrieked Jane and Alison together.

They were standing in front of their lockers taking

off their denim jackets, which were identical, each with silver stars sewn on the shoulders and down the front. Hillary was wearing the same jacket, too. Their mothers had bought them in the same store downtown even though they were rather expensive. It was such fun to dress alike, as if they were members of a select club.

"What kind of elves?" Alison asked suspiciously.

Hillary told them about the little yards. She described the stones bordering the yards and the neatness of it all.

"Are there gates?" Alison wanted to know.

"I don't think so," Hillary said.

"Are there chimneys? How about mailboxes?"

Hillary shook her head to both. "There's no furniture or anything inside. They're just, you know, little houses."

Alison shrugged. She and Jane looked at each other.

"I bet she made them herself," Jane said.

"Maybe she did," Hillary replied. "And maybe she didn't. You should come see."

"All right, we will," Alison said, pushing the sleeves of her sweater up her arms with two smart strokes. She was the best dressed of the three. Jane was brighter, though. Her mother was a lawyer. Why they had alighted upon baby-faced Hillary for the third in their group even Hillary didn't know. She was often awed by their sophisticated conversations.

"We'll come, but don't think we'll be fooled for a minute," Alison said.

Jane put on the sweet and innocent face that always meant something awful was coming. "I'd be interested to see if anything can live in that sickening backyard," she said to Hillary as the bell for class rang. "Besides Sara-Kate, that is."

She was not allowed to find out because Sara-Kate refused to have her. Sara-Kate refused to invite Alison either, though Hillary asked as nicely as possible. Jane and Alison waited out of sight in an empty classroom.

"But why?" Hillary begged. "You've got to let them come."

The thin girl shook her head and raised her voice slightly. "These elves are private people. They aren't for public display. You can come if you want, but not those two creeps."

"They're not creeps. We're friends," Hillary protested.

But Sara-Kate, who didn't have any friends, who spit at people when they made her mad and walked around all day in a pair of men's boots, only smiled faintly.

"Some friends!" she announced, in a voice that carried straight down the hall to Jane and Alison's furious ears.

"Sara-Kate Connolly is not a nice person," Alison said to the group when Hillary returned. "She gets in trouble a lot. Hillary should be careful of her."

Jane nodded. "Anyone can say she has an elf village in her backyard if she wants to. The point is,

· 11

where are the elves? I bet Sara-Kate is the only person who ever sees them."

"Nobody sees them," Hillary said. "Not even Sara-Kate. They go away when they hear people coming. Elves are very private persons. Sara-Kate said they used to be seen in the old days, but not now because there are too many people around and they're frightened. Elves haven't been seen for over a hundred years."

"If these elves are so real, why doesn't Sara-Kate want us to come look?" Jane inquired, casting a shrewd glance at Alison.

"Because they're fake," Alison answered without waiting for Hillary to reply. "Just like Sara-Kate."

"She's definitely not a person you want to trust," Jane agreed. She lowered her voice and drew the friends closer. "Do you remember that new bike she was riding to school last spring? Do you remember how she boasted about it and said she had a job on a paper route? Have you noticed how she isn't riding it anymore this fall?"

Alison nodded.

"What happened?" Hillary asked.

"She stole it," Jane whispered. "From a store downtown. Everybody knows. The police came to Sara-Kate's house and she was arrested. Only, she gave the bike back so nothing happened. They're watching her, though, in case she steals something else."

Hillary was shocked. "How awful!"

For the rest of the day she kept away from Sara-Kate. When she walked home from school, she saw

her thin shape in the distance and it looked dangerous suddenly. It looked like the shape of someone
who was bad, someone who lived in a bad house
and came from a bad family.

If magic had truly invaded Hillary's room, now
it slithered away again. It was gone by the time she
reached home that day, and Hillary was relieved.
She felt as if she had made a narrow escape and
laughed at herself for being so easily fooled. She
began to remember other incidents connected with
Sara-Kate Connolly. They were little things—a lost
pencil case, a series of small disappearances from the
art room, a mean note left in someone's desk. Taken
together, they added up to something larger in Hillary's mind.

"I do think it's best not to spend time down in
that yard," Mrs. Lenox said, approvingly, at dinner.
"Heaven knows what you might catch or step on."

Two days later, Hillary had put the elf village
almost completely out of her mind when Sara-Kate
appeared at her elbow in the hall at school. She
appeared so suddenly, and at such an odd time—all
the other fifth graders were at sports—that Hillary
jumped.

Sara-Kate leaned toward her and spoke in a high,
breathless voice.

"Where have you been? I thought you were coming again. The elves have built a playground. They
have a swimming pool and a Ferris wheel now." She
flung a string of hair over her shoulder and smiled

nervously. "You should come see," she told Hillary.

"A Ferris wheel!" In spite of herself, Hillary felt a jab of excitement. "How did they build that?"

"With Popsicle sticks and two bicycle wheels. It really goes around. The elves come out at night and play on it. Really and truly," said Sara-Kate, looking into Hillary's eyes. "I can tell it's been used in the morning."

Hillary glanced away, down at the floor, where she noticed that both of Sara-Kate's boots were newly speckled with mud. Her legs rose out of them, two raw, white stalks that disappeared under her skirt's ragged hem. She didn't seem to be wearing any socks at all. Half of Hillary was repelled. No one in the school was so badly dressed as Sara-Kate, or so mean and unhealthy-looking. And yet, another half was strangely tempted.

"Maybe I could come over this afternoon," she told the older girl. "Just for a minute, though. I've got a lot of things to do."

Sara-Kate's small eyes narrowed. "In that case, don't bother."

"I want to," Hillary said, "but my mother—"

"Who cares!" Sara-Kate interrupted. "Who cares about your stupid mother."

"She's afraid I'll catch poison ivy."

"Do I have poison ivy?" Sara-Kate extended one of her skinny arms for Hillary's inspection. "Is there one bit of poison ivy on me?"

Hillary shook her head. Sara-Kate's skin was pale,

14 ·

but unmarked. Her nails were cut short and her hands were clean.

"You won't catch poison ivy, but don't bother to come anyway," Sara-Kate said. "These elves don't like a lot of people looking at their stuff. They aren't show-offs like most of the creeps around here."

"It isn't that," Hillary tried to say, but Sara-Kate had turned her back. She began to walk away, and Hillary could see from the stiffness in her shoulders and the line of her chin that she was hurt.

"Wait a minute!" she called. "Wait! I forgot to ask you something."

But it was too late. Sara-Kate had passed beyond the limits of reasonable conversation.

"I know what it is," she sneered over her shoulder, "and I'm not answering. Even if I told you you wouldn't believe me. You wouldn't, would you?" she shouted at Hillary, while other people in the hall stopped to stare at her: at her clothes and her boots and her hair falling over her face. "You just wouldn't, none of you!" she shrieked, losing control in a way most unlike her. She began to run and hop along the hall in the strangest fashion, with knotted fists and flying feet. Like an elf, Hillary thought. Sara-Kate's face had turned bright red. She looked exactly like a tiny, silly, cartoon elf trying to run away fast and getting nowhere.

"Creeps!" Sara-Kate screamed, with her boots drumming into the floor.

Along the corridor, groups of students moved carefully out of her way.

Three ✒

At a little past three o'clock on the same afternoon, Hillary went through the hedge into the Connollys' backyard. She sneaked through, looking first right, then left, but whether she was afraid of being seen by Mrs. Lenox, or by Sara-Kate on the other side, or by the elusive elves themselves, she didn't know. She felt sure that she should not have come at all. Sara-Kate was too strange. Her house was too shabby. Hillary should have stayed home, safe in her kitchen. She should have read a book or baked a cake with her mother. There were ten other things she might have done, but, by the thinnest thread of enchantment, the elf village was drawing her.

How did it do that? What was its magic? Hillary could not shut out pictures of the curious houses that crept into her head. She could not forget the leaf roofs. The little well appeared and disappeared, wavered and vanished in her imagination in a most maddening way, like Alice's Cheshire Cat. The Ferris wheel was the strongest lure. She could not quite visualize how it would look, and came out her back door in a sort of trance from trying so hard to see.

Hillary's enchantment did not extend to Sara-

16 ·

Kate, however, and the closer she came to the Connollys' yard, the more she hoped that Sara-Kate would not be there. The voices of Jane and Alison came back to her. She heard their sensible warnings again, but softer now, muffled by some other power.

The boughs of the hemlock hedge presented Hillary with a thick, green curtain. Passing through, she was forced to raise her arms to protect her face, to close her eyes—and, for a moment, there was a frightening feeling of walking blind into a trap. But when she opened her eyes, only the Connollys' backyard came into view, as weedy and trash-strewn as ever. Away to the right, a figure slouched on a pile of wooden planks, looking more like the lone survivor of a wreck at sea than the violent, unpredictable girl it must be. Hillary shoved her hands into her pockets. She approached warily and had come within a few feet of the woodpile when Sara-Kate's head turned and her two tiny eyes flicked wide with surprise.

"You!" Sara-Kate exclaimed. Then she sprang to her feet, and whatever had been bowed or sad about her before vanished in an instant. She leapt off the pile of planks, landing exactly beside Hillary. Her boots made almost no sound hitting the ground. Sara-Kate gathered her long, straw-colored hair behind her head with a sweep of one hand and let it fall down her back. She grinned and hooked her thumbs into the waistband of her old skirt.

"Hi!" she said. "I thought you weren't coming."

"Well, I did," Hillary mumbled.

"The Ferris wheel's over there," Sara-Kate said, pointing.

They went to look right away. Hillary drew a deep breath. It was twice as big as she had expected, and constructed in such a complicated manner that one look told her it was the work of special hands. Two bicycle wheels without their rubber tires were suspended face to face above the ground on a metal rod that passed through the wheels' centers. The rod's tips rested on up-ended cinder blocks. The Popsicle sticks that Sara-Kate had mentioned were attached by strings to the wheels' outer rims and hung down horizontally to act as long seats. Numerous pieces of wire linked the spokes of the two wheels, so that when they turned, they turned as one and a marvelous pattern of wires and spokes was woven before the eye.

"How did it get here?" Hillary asked softly.

"It just was here, yesterday morning when I came out of the house," Sara-Kate replied, with such wonder in her own voice that Hillary had no doubt it was the truth. She looked at Sara-Kate with admiring eyes.

"I'm still not sure how the elves make it go," Sara-Kate went on. "I've gone over the whole thing and I can't find a motor anywhere. Probably they have some power or current that we don't know about. Naturally, we humans have to spin the wheel by hand."

She reached out as she spoke and spun the wheel

hard, transforming the wires and sticks into a series of gold and silver flickers.

"Now you try," she said to Hillary. She didn't mention her rage in the hall at school that morning. She treated Hillary respectfully, as if she were a special friend.

"It doesn't matter where you hold the wheel," Sara-Kate said in a most reasonable and helpful voice. "Just grab it anywhere and spin."

Hillary, who had never ridden on a Ferris wheel, leaned forward shyly and turned the wheel. It was very large, so large that it would have lifted a person far, far off the ground, she thought . . . if that person happened to be the size of an elf.

Hillary crouched beside the Ferris wheel. She made her eyes level with the highest seat and looked to see what an elf's view of the yard would be. There was the rusty white side of the washing machine rising through weeds like a mountain peak. There was the top of a tree stump appearing above the green jungle like the smoke stack of an ocean liner. The yard looked enormous from this vantage, and dense with greenery. In the distance, Sara-Kate's house towered over all, a rather frightening gray fortress.

"Have you ever wondered what it would feel like to be an elf?" Hillary asked Sara-Kate. "I mean, how would it feel to be so strange and little?"

Sara-Kate's eyes jumped to Hillary's face.

"What do you mean 'strange and little'?" she inquired sharply. "If you were an elf you wouldn't feel strange or little. You'd feel like a normal, healthy elf."

"Sorry," said Hillary, "I didn't mean—"

"Yes, you did," Sara-Kate interrupted. She pushed Hillary's hand away from the Ferris wheel and halted its spin with a single finger stabbed between the spokes. "I guess I'd better tell you something right now, before we go any further." Her voice was soft but forceful. "Nobody insults these elves and gets away with it. Not while I'm here. Nobody insults them, and nobody insults them by mistake, either," she added, seeing that Hillary was about to protest again. "Before you say anything, you've got to put yourself in the position of the elf. That way you don't make mistakes, okay?"

Hillary nodded. She leaned over and touched the Ferris wheel with the tip of her finger. She didn't want to get into another fight. The Ferris wheel was so wonderful, and besides, she could see how Sara-Kate might be right, especially if there happened to be an elf nearby listening to their conversation.

An elf nearby listening? Even as she thought this, Hillary felt an odd sensation on the back of her neck. It was as if a small hand had passed between her skin and the collar of her jacket. She glanced over her shoulder at a bush behind her.

"I know. I felt it, too," Sara-Kate said quietly, following the direction of Hillary's eyes into the bush. "I have feelings like that all the time here."

"You do? Do you think it means . . . ?"

"It's better not to talk about it," Sara-Kate whispered. "It's better to keep on doing things and not look.

"Come and see the elves' new pool!" she shouted suddenly, in a voice clearly intended for invisible ears. "It's over here! Follow me!"

Then: "Quick, come on," she whispered, and rose swiftly from her knees. Hillary rose, too, and the two girls scampered away, feeling such a pressure of elfin eyes at their backs it seemed almost that they were propelled across the yard.

That night, in bed, Hillary put her face against the window and tried to look through the dark. Now the elves were in their village. Now, if she could only see, they were walking in their front yards, sitting in their houses, talking in tiny voices among themselves. She could feel them out and about, mysterious little beings scurrying through Sara-Kate's backyard, over the broken glass, around the washing machine. Was the Ferris wheel turning? She peered into the blackness.

"Elves are almost invisible," Sara-Kate had said. "It isn't that they hide so much as that they decide not to be seen."

"But they have been seen. Some people have seen them," Hillary said. "You said so yourself."

"Right," answered Sara-Kate. "Some people. The right people. People they can trust."

"Do you think the elves could ever learn to trust us?" Hillary asked.

"That's what I'm hoping," Sara-Kate said. "But

don't count on it," she'd added immediately. "It takes a lot for an elf to get his trust working."

"But why?" Hillary had asked. "We wouldn't hurt them."

"But a lot of people have," Sara-Kate replied.

Four

Sara-Kate had said "put yourself in the position of the elf." During the next week, Hillary found herself slipping into that position frequently and with remarkable ease. It did not feel odd or unnatural at all, especially with Sara-Kate hovering watchfully nearby, whispering, explaining, drawing upon an apparently endless supply of information about elves.

She knew everything that could be known about them, it seemed to Hillary, whose eyes now often followed the older girl's unusual figure at school, along halls, into classrooms. She began to wait for her near the cafeteria at lunchtime, to sit with her if Sara-Kate was willing, though she more often passed Hillary by and went to occupy a chair away by herself.

Jane Webster and Alison Mancini watched their friend's new attachment with alarm. They took Hillary aside and tried to warn her.

"What is wrong with you?" Jane hissed one day outside the lunchroom, where she had come across Hillary standing rather pathetically against the door frame. "Sara-Kate Connolly is not a good person. She's out to trick you and everybody knows it. Everybody keeps telling you to watch out, to stay away from her. But do you listen? No! You're over at her house every afternoon. You're walking home from school with her every day. And why are you standing around here waiting? She never sits with you anyway."

"Sometimes she does," Hillary replied.

Jane sighed and tried another approach.

"Have you seen what Sara-Kate eats for lunch?" she asked. "She brings white mush from home and pours sugar on top. White mush! Can you believe it?" Jane's eyes widened in horror.

"It's only Cream of Wheat cereal," Hillary answered. "Sara-Kate has a delicate stomach. She can't eat hamburgers and pizza and things like that. She cooks the Cream of Wheat herself in the morning and puts it in a thermos. That way, it's hot for lunch. She told me."

"Do you know that Sara-Kate's father is a criminal?" Alison asked Hillary later that day. "He's in prison for armed robbery and will probably be there for a long time. A friend of my mother's told her."

But Hillary only smiled. "He's not in prison, he's in Sarasota, Florida. Sara-Kate said so," she replied with such honest conviction that Alison fled to Jane in a fright.

"It's as if Sara-Kate has put a spell on her!" she whispered to her friend. "Hillary believes everything she says. Everything!"

Spell or no spell, magic or none, Hillary was getting more attached to the Connollys' backyard with

24 ·

each passing day. There was a lot of work to do around an elf village, she discovered. She could not just sit still and watch because even as she looked, a leaf roof would blow off and she'd have to run after it to bring it back. Or a line of pebbles would become crooked and need to be rearranged. The elves appreciated this kind of light repairwork. But they would not stand for too much meddling with their village, as Hillary soon discovered.

During a rainstorm, two of the tiny houses entirely collapsed. Hillary and Sara-Kate found them the next day. Hillary kneeled right down to begin putting the structures back together, but Sara-Kate jumped in front of her and grabbed her wrists.

"Don't touch!" she yelled. "These are elf houses and only elves can build them right. People don't know how!"

Hillary snatched her hands away angrily. "You never told me that," she said. "How am I supposed to know things you haven't even told me yet?"

"Well, it's obvious, isn't it?" Sara-Kate spat back. But then, seeing Hillary's expression, she said in a kinder voice: "It's all right. Don't worry. We can help the elves with little things. We can leave presents for them. They would like that."

"What kind of presents?" Hillard asked.

"Food!" announced Sara-Kate with a broad smile. "Elves love to eat."

Who would have thought there could be so much elf food in that brambly, neglected backyard? ("I guess that's another reason the elves came here," Hillary said to herself.)

In the brambles grew bright red berries.

"Elf apples," explained Sara-Kate, picking them off with her thin fingers.

Out of the mud appeared pure white mushrooms.

"Poisonous to humans," Sara-Kate said. "But to elves they are soft and sweet as cake."

There were also sticky green pods that contained tiny white seeds.

"Elf salt?" asked Hillary.

"Right," said Sara-Kate.

And there were blackberries and little pink flowers in the underbrush that Hillary's father would have called weed flowers. There were no weed flowers left in Hillary's backyard, and no place was muddy enough to grow mushrooms.

"You've got a perfect yard for elves," Hillary said to Sara-Kate wistfully. "Nobody in the whole neighborhood has a yard even close to this."

Sara-Kate pushed her nose up in the air. She said, "I know. I've been knowing it for a long time."

"Do elves eat regular flowers?" Hillary inquired. "We've got a whole lot of pretty ones growing in our yard."

"They hate them," Sara-Kate answered. "Regular flowers are poisonous to elves."

"I thought so," Hillary muttered. "I've been noticing that about elves."

"What?" Sara-Kate said.

"That what's poisonous to people is healthy for elves. And what people think is pretty is not at all what elves like to live near."

"You're getting to understand elves pretty well," Sara-Kate allowed. Then she found two caps off the tops of acorns that, filled with water, looked exactly like the sort of cups elves would drink out of. These they left on a leaf near the village, surrounded by all the food.

"Will it be safe?" Hillary asked, standing back to admire the banquet. "Shouldn't we cover the food with something? A dog could come along and wreck this in a minute. And it's getting so cold and windy out here," she added.

The afternoons had grown progressively chillier during the week. September was nearly over and there was a feeling of changing seasons in the air. On this particular afternoon, the wind had a nasty bite to it that now caused Hillary to turn up the collar of her thin jean jacket. Sara-Kate looked at her and shrugged.

"Don't worry," she said. "I always check things before I go to bed. And then I come look in the morning before school. It takes a lot of work to keep elves, but it's worth it."

"What will they do in winter?" Hillary asked. "They'll get pretty cold out here." Her own feet felt icy suddenly, and looking down she saw that her sneakers had gotten rather muddy.

"Elves don't get cold," Sara-Kate said.

"Everybody gets cold."

"Not elves," Sara-Kate said proudly. "They like being outside. They have thick skins. They never go inside until they have to. Houses are too hot for elves. They can't breathe right."

"Then why did they bother to build all these houses?" Hillary inquired a little sharply. She could not quite believe it about the elves' skins. Even furry animals got cold, she knew. In winter, they burrowed into caves and nests and went to sleep.

"Why don't elves just live in trees or underground like other animals?" Hillary asked Sara-Kate. "It would be so much easier for them."

Sara-Kate shook her head. "You can be pretty stupid sometimes," she scoffed. "The reason they build houses is to have a village so they can live together, of course. Elves keep together. If they lived in trees or holes, they'd be all scattered out." She squinted at Hillary. "And anyway, elves aren't animals," she added. "For one thing, they're about a hundred times smarter than any animal. They're about ten times smarter than most people and about twice as smart as a human genius."

Sara-Kate stopped suddenly and looked around toward her house. There, some signal invisible to Hillary must have caught her eye, because she began to walk rapidly in the direction of the back door.

"You've got to go home now," she told Hillary over her shoulder. "My mother wants me to come in."

When Hillary stared after her in surprise, Sara-Kate flung herself around again and bellowed, "Go on! Get going!"

She disappeared into her house with a slam of the door.

Hillary sighed. She glanced a last time at the elves' banquet to see that all was in order. She shivered. The wind had stopped coming in puffs and now blew in one long, cold stream.

All of a sudden, one of the leaf roofs came detached at one side and was blown up straight by the wind. It was a deep red color and had the curious look of a hand, fingers and thumb outstretched, waving at her. It appeared so real that Hillary wondered for a second if the elves were behind it, playing a game with her.

She smiled at this thought, and had bent down to fix the leaf when another flutter caused her to straighten up quickly and look toward Sara-Kate's house.

She saw right away what it was. A shade in one of the upstairs windows had been flicked up, and now, as she watched, a thin face rose where the shade had been and stared down at her with wide eyes. For a moment, Hillary stared back. Then she stepped away and ran for the hole in the hedge that led to her own yard.

When she reached it, she looked again but the face had vanished. The shade was already drawn back into place.

"Silly," she murmured. "It was only Sara-Kate's mother."

But the face had not looked as if it belonged to a mother, any mother. It had been too white and too thin. Too frightening.

Five

The weather turned warm again. Indian summer, Hillary's mother called it, and she was happy because "I wasn't ready to close all the doors and windows yet. It's so lovely to have the fresh air! And you'll need a new winter coat," she said, measuring Hillary's ever-lengthening legs with her eyes. "Now we'll have time to shop around for the best one."

Hillary's father was happy, too. On most evenings, he left his briefcase in the hall, changed from his business suit into a pair of jeans, and went out to work in the yard. He weeded and watered and planted daffodil bulbs, whistling to himself and addressing remarks to the garden at large.

"Hot enough for you?" he'd ask if the day was a warm one. Or, after a heavy rain: "Well, we had a regular drowning party out here, I see."

"You are a nuisance," he said one evening to a group of fall roses that were beginning to shed their petals. "I just raked you out yesterday and here you've gone and dropped your underwear all over the place again."

From the hedge at the bottom of the yard, a series of muffled snorts erupted. Hillary and Sara-Kate had been crouched behind the bushes building an elf bridge across a boggy section of ground. Now they looked at each other and grinned.

"Has your father always been crazy or is this a recent problem?" Sara-Kate asked Hillary, gravely.

This brought on a second round of snorts, and a number of giggles and coughs besides.

"Everything all right down there?" Mr. Lenox called unhelpfully from the garden. It made them crack up again.

Everything was all right, of course. It couldn't have been better, in fact. Strange as it might seem, Hillary and Sara-Kate were putting together a sort of friendship. They met in Sara-Kate's backyard every afternoon to work on the elf village. Hillary's mother disapproved, but she hadn't actually forbidden the visits, so Hillary was able to slip through the hedge with the understanding that no questions would be asked.

"If you could tell me which is the poison ivy, then I can keep away from it," she'd said to Sara-Kate at the end of the first week, during which she had hardly dared venture farther than the village and the Ferris wheel. "My mother's just waiting for me to catch something down here so she can stop me from coming."

"Don't think I don't know," Sara-Kate said. She took Hillary around the yard and pointed out the worst patches, and she showed her how to recognize the poison ivy plant, with its distinctive three-leaf cluster and green sheen.

"I used to get poison ivy all the time when I was little," she said. "Then I got wise."

"Why doesn't your mother do something about it?" Hillary asked. "My father said she could get a spray to kill it if she wanted to."

"Well, she doesn't want to," Sara-Kate answered quickly, in a voice that told Hillary not to go on with the subject. It wasn't the only subject that Sara-Kate wouldn't discuss. There were many others. She was always backing out of conversations, pulling up short, telling Hillary to mind her own business. Sometimes Sara-Kate simply turned her back and walked off without explanation, as she'd done on the day when Hillary had seen the face in the window. Being friends with Sara-Kate was a complicated business. But sometimes . . .

Behind the hedge, Hillary looked over at Sara-Kate, who was still laughing about the roses' underwear, and thought what a nice person she was when she let herself relax. She was really no different from anyone else, Hillary decided. She even managed to look rather pretty at times, after one got to know her and could ignore her boots and the strange clothes she wore.

The project underway at the elf village that afternoon was the construction of a network of roads or paths leading to different parts of the yard. The elves were a quick, energetic people who needed to be able to move around easily, Sara-Kate had explained. But they were not, by nature, roadbuilders, preferring to follow the trails made by other animals. There were no animals living in Sara-Kate's backyard, just a few squirrels whose roadways ran overhead, along the branches of two ragged trees.

"So we must take over the job of making roads," Sara-Kate informed Hillary. "And what I was thinking was, we could make a really good system. First we could design a plan for where the roads should go, and then we could carry it out."

"If you want me to make a map, I'm good at that," Hillary replied. "I studied maps in school last year. I know about scales and ledgers and things."

"Good," Sara-Kate said.

So, as official map-maker, Hillary had walked the boundaries of Sara-Kate's yard with a pad and pencil. She had marked the position of the overturned washing machine, of large bushes, trees, tree stumps, and of such smaller landmarks as derelict tires, rusty pipes, oil cans, and ash heaps. She noted the placement of the back porch and discovered the swampy area near the hedge, which was too soft to support a road and might be dangerous to a small being like an elf.

While she walked, watching her step for fear of broken glass or poison ivy, Hillary kept half an eye on the blank, shade-drawn windows of Sara-Kate's house, especially the upstairs windows. For there, at times, some tiny movement, an almost imperceptible flutter, seemed to catch her attention, and she would whirl around to look with a leaping heart.

But whether it was her imagination or a reflection trick of the window glass, she could never positively identify what had moved, or where. And she did not dare ask Sara-Kate about such things. She would only have laughed or become angry. She might even have ordered Hillary out of the yard for good, a possibility that worried Hillary far more than the feeling that she was being secretly observed.

Short of being ordered away, Hillary wouldn't have stopped coming to the yard for anything. The place fascinated her, and she liked the idea that she was beginning to know its parts: the stumps, the rocks, the junk heaps, and the hidden places where the tiny weed flowers bloomed. She liked knowing how to get around and between obstacles—how to steer clear of thorn bushes, for instance, by walking the trunk of a fallen tree. She had a feeling of belonging to the yard, even of owning it a little.

The Lenoxes' tidy yard belonged to her father and mother, Hillary saw. It was under their order and grew according to their laws. But Sara-Kate's yard was wild and free, and that was how Hillary felt there, tramping among the bushes, poking into shadowy dens. Anything might happen in Sara-Kate's backyard. For that matter, anything *was* happening.

The elves might keep themselves hidden. They might even be invisible to the human eye, as Sara-Kate believed. But everywhere, everywhere! there was evidence of their small, exotic lives.

A cache of acorn cups would turn up in the hollow of a tree root, leading Hillary to imagine an industrious band of little workers moving and storing goods throughout the garden.

Several odd, circular dirt clearings appeared in the weedy underbrush. The earth of these places was packed firm and level, and, though no footprints or marks showed there, it seemed obvious that these were meeting areas of some sort.

But why? What did the elves discuss? What language did they speak? Did they talk in words, or did communication take place through some form of silent sign language? After all, there was never a mutter or a cough in the yard, not during the day and not at night either, according to Sara-Kate. There was only the sound of the wind moving through bushes and tree boughs.

"If I were an elf and wanted to speak so that no one would hear me, I would make up a language that sounded like ordinary natural sounds," Sara-Kate had said.

"What sounds?" Hillary asked.

"Be quiet and listen for a minute." They had stood still until the tiny noises within the boundaries of the yard began to distinguish themselves from the louder noises of the town: from traffic passing in the street, the shriek of brakes, the cries of children and yapping of dogs.

Underneath these town noises, Hillary heard tiny chirps and squeaks. She heard a faint vibration or buzz coming through the air. She heard pecking and tapping noises, little creaks and scratches, groans and gushes, clickings and drippings. Finally, she heard the clear notes of a bird, very close by.

"Is that an elf talking?" she whispered in amazement.

Sara-Kate shrugged. "It could be. Who can tell?"

Hillary had nodded. It might very well be. But of course, with elves it was impossible to be sure of anything. One might suppose that certain things were happening, that the elves acted in a certain way, but who could really tell? Facts can be understood differently, they can add up to different answers depending on how they are viewed.

This uncertainty about the elves had come home to Hillary after a particularly interesting conversation with Sara-Kate one day, a conversation about the strange little pool the elves had made for themselves out of an old tin pan sunk in the ground near the Ferris wheel.

At night in the dark, or perhaps in faint moonlight, the elves bathed in the pool. Hillary knew they bathed because she saw fresh seeds and half-eaten berries near the water's edge. She saw small, square pieces of wood floating in the water, along with tiny, yellow leaves that seemed to have no source within the yard.

What purpose did the wooden pieces serve? Hillary wondered. And what about the leaves?

After some thought, she had come up with an answer. The wooden squares were rafts upon which the elves sat or lay, and the leaves were washcloths, because the elves must have some means of washing, mustn't they? It seemed most likely.

But Sara-Kate had laughed when she heard Hillary's ideas.

"Why do you think that these elves are anything like you?" she asked. "You play with rafts in pools and you use washcloths, so you think elves must, too. But maybe elves aren't like you at all. Maybe they're so different that nothing they do is anything like what you do. Maybe they've never even seen a washcloth and these leaves are for something else, for collecting starlight, say."

"Collecting starlight! Why?"

"Maybe for energy to run things, the way we use the sun for solar energy."

"To run what?"

"Well, there's the Ferris wheel. Suppose the wood pieces aren't water rafts but power rafts. Maybe the swimming pool isn't for swimming at all. It could be a power center that collects energy and stores it for future use. I'm not saying this is true," Sara-Kate added hastily. "I'm just trying to show you what's possible."

Hillary was astounded. Starlight collectors! She would never have imagined such a thing and she looked with respect at Sara-Kate. Not that she was convinced about the power rafts, but she saw that Sara-Kate was right in principle. She must not take anything for granted when it came to an unknown like elves. She must just watch and wait and hope that they would reveal themselves more clearly.

In the meantime, the road and bridge project went forward in the afternoons until most parts of the yard had been linked with the elf village. But Indian summer soon passed and the days grew shorter and colder, making it less pleasant to work outside.

Less pleasant especially for Sara-Kate, Hillary thought. She stubbornly refused to put on any kind of coat. While Hillary arrived wearing sweaters and jackets and finally a new quilted parka, Sara-Kate went on working in her same blue sweatshirt, which by this time was looking rather ragged around the cuffs.

"Aren't you cold?" Hillary asked her. On some afternoons, there was a thin layer of ice on the elves' pool.

"No," Sara-Kate answered.

"Are you trying to act like an elf and have thick skin?" Hillary said, teasing a little.

Sara-Kate had turned on her in one of her unpredictable bursts of fury.

"For your information, I'm not trying to act like anything. I happen to be like an elf, that's all. I don't get cold. If you think that's weird, why don't you go away and tell your stupid friends about it."

But, of course, Hillary chose not to go away.

Six

In all the time that Hillary had been going to Sara-Kate's backyard to work on the elf village, Sara-Kate had never once invited her inside her house.

Never once had she offered Hillary a soda or a snack. She had never asked her in to see her room, or to watch television. When it rained, Hillary went home. When she got hungry, Hillary went back to her own kitchen.

"Do you want something?" she'd ask Sara-Kate. "We have cupcakes." Or she'd offer an apple, popcorn, lemonade, for Mrs. Lenox believed in a well-stocked larder and kept many delicious snacks on her shelves.

But Sara-Kate never wanted anything. Elf banquets were one thing. For herself, she did not seem interested in food. She didn't like to watch Hillary eating either, or so it seemed, for she would turn away and go off to another part of the yard while Hillary nibbled around the edges of a raspberry fruit pie or a chocolate-chip cookie she had brought from home. It was as if Sara-Kate found the sight of this sort of food disgusting.

On the other hand, Hillary occasionally caught

her tossing a handful of yard berries into her mouth when she thought no one was looking. And she often munched on mint leaves from a patch of mint grown wild near the back porch.

"Elves chew mint like gum," she told Hillary once.

Hillary watched Sara-Kate chewing and guessed that her odd tastes were another instance of her being "like an elf."

"If you ask me, Sara-Kate *is* an elf," Jane said in a nasty voice to Hillary at school. "She's a mean, dirty little elf who's put a spell on you and is going to get you in trouble, wait and see."

Jane and Alison were angry at Hillary and she was angry at them.

"You think you're so great because a big fifth grader is your friend," Alison hissed. "But nobody in the whole school likes Sara-Kate. Nobody else wants to hang around her horrible old house. There's not even any furniture in there, did you know? Somebody sneaked up and looked through the window one day. There aren't any chairs or anything."

"Sara-Kate is your friend because she can't get anybody else," Jane added. "She got you by telling you a lot of lies about elves. Why are you so stupid, Hillary?"

"Shut up," Hillary replied. "Sara-Kate isn't lying. And she has beautiful furniture. Whoever said she didn't is the one who's lying. I've been in her house hundreds of times so I should know."

Not only had Hillary never seen the inside of Sara-Kate's house, she had not seen Sara-Kate's mother since the day she'd stared out the second floor window for that brief moment. Mrs. Connolly never came out of the house. She never came to the door to call Hillary inside. She never made a sound.

And yet, she was there. Hillary knew it. Besides the way the shades moved at times—as if someone were crouched behind them watching the activity in the yard—there were the errands that Sara-Kate was constantly being sent to do.

Hillary had even begun to accompany her: to the drugstore to pick up a prescription; to the post office; to the little grocery store two blocks away for a bunch of carrots, a half gallon of milk, a box of Saltines. Hillary went secretly, of course. Her parents would never have allowed her to walk around the town, which was more like a small city in the downtown sections. They would have worried about traffic on the busy streets, about the beckoning finger of a stranger. They would have worried that Hillary would become lost, or get bitten by one of the homeless dogs that roamed the trashy alleys, or fall into a hole and lie there unconscious and ignored.

Apparently Sara-Kate's mother worried about none of these possibilities. Hillary wondered if she knew where Sara-Kate was half the time. And naturally Sara-Kata didn't worry. Perhaps she'd never been told about the terrible things that can happen to a child alone on the streets. She went to the laundromat to wash clothes, to the hardware store to "make a payment," to the bank to cash a check.

Hillary watched Sara-Kate's small figure transact these grown-up pieces of business with increasing amazement. After all, Sara-Kate was no bigger than Hillary, and though she was two years older, she did not look old enough to be so effective in the adult world. But effective she was, although Hillary once heard her questioned by a woman in the business office of the telephone company, where she had gone to restart telephone service that had been shut off at her house.

"Where is your mother?" the woman inquired sharply. "She should be handling this."

"She's sick, so she had to send me," Sara-Kate replied quickly. The telephone official regarded her doubtfully, but she accepted the pile of money that Sara-Kate placed on her desk.

"How do you know what to do?" Hillary asked Sara-Kate later. "Do you do everything for your mother?"

"Not everything. Just what she tells me."

So Sara-Kate's mother was sick. That scrap of information was among the few that Hillary managed to glean during her month of visiting the Connollys' backyard. Sara-Kate almost never spoke about herself. She never told stories about her family. If Hillary forgot and questioned her too closely on some personal matter, Sara-Kate snapped at her. Or she was silent, as if she had not heard.

"What sickness does your mother have?"

Silence.

"Are you the only child in your family, like me?"

Silence.

"Why was your telephone shut off?"

Silence.

Only about the elf village and the elves themselves did Sara-Kate became talkative and open. In fact, as time went on, the mysterious world of the elves became clearer and clearer in Hillary's mind from Sara-Kate's telling about it, and Hillary could almost see the pale, quick-as-a-wink faces peer out from the underbrush.

Perhaps she had seen one, or rather part of one. As Sara-Kate explained, the elfin trick of invisibility depended to a great extent on their never appearing whole before humans. Sara-Kate herself saw bits and pieces of elves everywhere in the yard: a flash of arm, a pointed foot, an eye, wheaten hair blowing in the wind like yellow grass.

"It isn't where you look for elves so much as how you look," she advised one day when Hillary had despaired of ever seeing the little people. "You've got to train yourself to notice details. You can't just stomp around the place expecting to be shown things. Go slowly and quietly, and look deep."

So Hillary went slowly and quietly. She began to look deep into the bushes of Sara-Kate's yard, deep into piles of leaves, into hedges. The yard seemed more open now because the trees and bushes had shed their leaves. She saw holes, hollows, and stumps that hadn't been visible under all the foliage.

"I think I just saw an elbow!" she would call to Sara-Kate, who would turn toward her skeptically.

"Are you sure?" she'd ask. "Are you really sure?"

"Yes!" Hillary would cry, but she wasn't. She was never sure what she had seen. It was maddening.

Hillary found herself bringing her new noticing eyes back home with her to her own yard. She noticed, for instance, how the ivy climbing her father's birdbath turned brittle, then brown, and began to lose its grip on the fluted stem.

She saw an abandoned nest in the bare branches of the apple tree. She noticed how evening came earlier and earlier, until it was completely dark when her father returned home from his office and there was no time for him to work in his garden. But then again, she saw that frost had killed the flowers and the grass had stopped growing so there was no reason for him to go out there anyway.

At the dinner table, Mr. Lenox looked tired and talked about people who weren't doing their jobs at work. When Hillary's fork dropped on her plate with a crash by mistake, he jumped and scowled at her. Looking deep, Hillary thought she knew why.

"He misses his garden," she whispered to herself.

"You're so quiet these days I hardly know when you're in the house or out," Hillary's mother said to her. "Would you like to invite Jane or Alison over for the night this weekend? Or how about both!" Mrs. Lenox offered, a little wildly, because Hillary had been worrying her lately.

"You're spending far too much time in the Connollys' yard," she might have added, but didn't since the subject was an embarrassing one. Mrs. Lenox disliked the Connollys' shabbiness. She was nervous about the disorder lurking just beyond the hedge. It nibbled at the edges of her own well-kept yard. But how could she speak to Hillary about the unsuitableness of such a house, such a family? She couldn't. Hillary looked deep and heard her anyway.

"I can't come here every day anymore," she told Sara-Kate. "My mother is noticing it too much."

Sara-Kate shrugged. "Do what you want," she replied, as if she didn't care. Then her small eyes scurried to Hillary's face to see what could be read there.

"I want to come," Hillary said, meeting her gaze. "It isn't me that wants to stay away. It's my mother."

"Sure, I know," Sara-Kate answered. "You don't have to say it. I already know."

Hillary often wondered about Sara-Kate's life inside her gloomy house. In early November she finally got a chance to see. Even then, it wasn't by invitation but because no one would come to answer the door. She'd knocked and knocked. She'd stepped back near the elf village and tried calling up to the second floor.

"Sara-Kate!" she bellowed. "It's me! What's wrong?"

Not one sound came from anywhere inside and an odd little fright had crept into Hillary. The day was icy and dark. She jumped up and down to keep her feet from freezing. The yard lay around her, hunched under the cold. Bushes whose languid shapes she had come to know well in warmer weather now posed in awkward positions, their scrawny limbs angled and bent.

"Sara-Kate, where are you!"

Sara-Kate had not come to school for three days. Hillary couldn't remember her ever missing before. Odder still, she had not been outside working on the elf village when Hillary had come through the hedge to visit on Monday afternoon. That day, Hillary had gone home quietly. And she had gone home again on Tuesday when Sara-Kate had not appeared in the yard, but with uneasy feelings.

"Are the Connollys away?" she asked her mother, a ridiculous question since Mrs. Lenox had no more knowledge of the Connollys' comings and goings than of a family in Outer Mongolia.

"I thought we might go to the shoe store this afternoon to find something pretty for you to wear at Thanksgiving," her mother said, presenting such a happy prospect that Hillary forgot to worry about Sara-Kate for the rest of the day.

But on Wednesday there was again no sign of her and so, in the afternoon, Hillary approached the Connollys' back door and bravely knocked. Her fist made such a tiny, hollow noise—it was as if the house were completely empty—that she began to pound with the flat of her hand. There was no doorbell and no knocker.

"Sara-Kate!" Clouds of breath exploded from her

mouth. Behind her the yard was silent, listening. Or perhaps it didn't listen. Hillary looked over her shoulder. Perhaps it was as empty as the house. Not a twig moved. Not a bird chirped. In the elf village, many cottages were falling apart. The Ferris wheel stood grandly over the ground, but brush and debris had blown into its wires and the Popsicle-stick seats were tossed and tangled.

Hillary put her mittened hand on the doorknob and turned. The door came open.

"Sara-Kate?" she asked softly of the darkness within. Then, since there was no answer here either, she stepped forward across the threshold.

Seven

If, by some charm, Hillary had been shrunk to a height of three inches and escorted through the door of one of Sara-Kate's elf houses, it would have seemed no stranger than the place she entered now. Certainly, whatever peculiarities an elf house might have had—stars adrift near the ceiling, hypnotized moths on the walls?—it would have been more welcoming. This room was cold as ice, and dim. Hillary could distinguish little at first, just a heap of mysteriously shaped objects rising from the floor in front of her.

Gradually, as her eyes adjusted, she made out a table and then two bulky chairs from the heap. A cardboard box, a floor lamp, a stool, and a low bookcase came into view. They were set in a sort of circle in the middle of the room and, at their center, most incongruous of all, was a large white stove. But what a strange-looking stove. Hillary took a step nearer. The oven had no door, only a cavernous mouth where the door should have been. And the burners on top were gone, leaving behind four empty craters.

An electric fan was perched on a bureau positioned next to the stove. The bureau's drawers had been removed, however, to serve other functions. Hillary caught sight of one in the corner being used as a container for tools. Another, turned upside-down over a stool, had become a table top. Meanwhile, the bureau's drawer slots had become storage holes for cooking pans, jars, utensils, newspapers, and other items unidentifiable in the poor light.

In fact, everything in the room seemed to have been dismantled, rearranged, and transformed into something else, Hillary noticed during the full minute she stood gazing about just inside the back door. She saw that the room had been an ordinary kitchen once. There was a refrigerator against the far wall, and a sink and faucet were visible beneath a window whose torn shade allowed outside light to enter. For the rest, everything not actually attached to the wall or the floor (the radiator was still in place for instance) had been uprooted and dragged to the room's center, where it was installed around the stove to make another room. A room within a room.

But now, even this odd inner room seemed to have fallen into disuse. Hillary walked around it cautiously, examining its different sides. The two shabby armchairs, worn to cotton stuffing in the seats, had been pushed so close to the stove's open mouth that the people sitting in them must have baked their legs if the oven had been turned on. Hillary drew her mittened hand along a chair back. A fine dust rose.

Sara-Kate and her mother had gone away, this much was clear. The house was shut down. The heat was turned off. They had gone to visit friends or perhaps they were in Sarasota, Florida, with Sara-

Kate's father. They had left quickly, without telling anyone, and they had forgotten to lock the back door. Not that there was anything here to steal. Never had Hillary seen a place so stripped of the basic comforts.

She finished circling the inner room and arrived back near the door. But instead of opening it and going home, she walked across the floor to a doorway on the other side that led to another room. She could understand now why Sara-Kate had never wanted to invite her in.

The second room was completely empty and made her remember Alison's report on the state of the house. The walls were bare, the windows had no curtains, and when Hillary walked in, the sounds her feet made on the hardwood floor were as loud as hammer blows. The smallest sniff or rustle was magnified by the emptiness into an alarming hiss. She examined the room's yellowed window shades, all drawn but one that had fallen askew. She noticed the dust lying black on the sills, the cobwebs trailing from the ceiling. This was not a room recently cleared of its contents. This room had lain empty for months, maybe years, accumulating mold and rot and insect bodies, of which there were hundreds on the floor, Hillary saw. She stepped to one side with distaste. Had this been a dining room once? An ancient film of flowered wallpaper coated the walls.

She was near another doorway now, and through it she saw the shadowy lines of a staircase leading to the second floor. She did not want to go there. The cold in Sara-Kate's house was intense. Hillary pushed her mittened hands into her parka pockets and hunched her shoulders. Her nose and cheeks were icy. She turned back to the kitchen and had already taken several steps in the direction of the back door when she stopped. A faint noise reached her ears. It was just barely audible through her wooly hat. She took her hands from her pockets and pulled the hat off.

The sound was coming from somewhere above her, a series of little thumps as if something were being knocked or rolled upon the floor. The noises stopped as she listened, then started again a few seconds later. Hillary's heart jumped.

Perhaps she had been frightened all along and not noticed. The house had seemed more odd than threatening, more closed up and left behind than eerie. But maybe, all the time, protective antennae in Hillary had been picking up warning signals: a wisp of a smell, a current of air, a sigh, an echo. For suddenly she knew she was not alone in Sara-Kate's house. Someone was here with her.

Above, the sounds stopped again. Hillary held her breath and stood absolutely still. She calculated the direction of the noises, and was looking up, toward the back rear corner of the old kitchen, when they began again—a rolling, a knocking—just where she was staring. The sounds were coming through the ceiling from the second floor.

Hillary knew she should run. She thought of racing for the back door, flinging it open, crossing Sara-Kate's yard in three strides and without looking back crashing through the hedge into her own safe garden. Sara-Kate's house was too cold and dark for any usual sort of being to be living in it. But in that case, Hillary reasoned in a rush of thoughts, the source of the noises must be some thing or things unusual. It must be some thing or things with fur coats or hot blood for staying warm. Or perhaps possessing thick skins?

Hillary didn't move. She stared at the ceiling. She remembered how empty the yard had seemed these last three days. She thought of the elf village, whose houses were beginning to collapse, whose leaf roofs were blown away. Their fragile construction had never been intended for winter living. She recalled Sara-Kate's words: "Elves never go inside until they have to. . . ." *Until they have to.* And with Sara-Kate and her mother away, the house must have seemed the most logical place to go.

Over her head, the little roly-knocking noises had stopped again. In the silence, Hillary found herself turning quietly and tiptoeing through the door into the vacant dining room, across its dusty floor, and beyond another door. She came to the foot of the dark staircase in a hall so bitterly cold that it seemed to her the ice heart of the whole frozen house.

Hillary crept upward, stair by stair. She moved slowly and made little noise. The stair boards are too cold to squeak, she thought. She took shallow breaths to quiet the sound of her breathing and was lightheaded when she reached the top. Here, she paused, holding onto the banister.

A dark corridor swept past her. Away to the left, she heard the noises begin again, but for a little while she stayed where she was, drawing in gulps of icy black air. She was not frightened now, but filled with anticipation. Her fear was that the elves would hear her coming with their quick, pointed ears. She was sure they would perform some vanishing trick if they suspected her presence. Or they would run or fly away. Her plan was to burst suddenly into the room, to catch them unaware for a moment. And in that moment, Hillary would see an elf whole at last, which even Sara-Kate had never done.

Hillary leaned on the banister and tried to imagine what the elves would look like. She prepared herself for their small bodies, for their elf-made clothes— little hats and coats, little shoes with curly tips. For some reason, she had imagined them dressed in a bright spring green, but now she realized this might be another case of coming to the wrong conclusions. More likely, elves change color with the seasons, like chameleons, Hillary thought. She smiled. She was ready for anything. Sara-Kate's elves might have pink hair and purple eyes for all she cared. She was ready.

At last Hillary's breath steadied and she began to feel her way down the corridor. She moved with agonizing slowness, keeping one hand on the wall for balance. The noises had continued on and off during the time she had rested, but there was no need to listen to discover which room they came from. On the right-hand side, about halfway down the hall, light streamed beneath a closed door. It was not a pale, mysterious light, but a fiery yellow one. It poured through the keyhole as well, and through the cracks at the door's edges. To Hillary, approaching down the dark hallway, there seemed a great pressure of light upon this door, so that it bulged with the effort of holding the brightness in.

Or was magic what was being contained? Hillary came forward and stood before the throbbing door as if drawn by a magnet. She stared at the plain brown wood. Her mittened hand rose to encircle the doorknob and turn it slowly. The roly-knocking noises continued while the door came silently open and the room was revealed in a rush of heat and light.

Hillary did not believe what she saw. She looked again and again and could not understand. Then she understood. The elves had played a trick on her after all. They had sensed her coming at the final moment and vanished through the walls. They had substituted another scene to confuse her.

Before Hillary's dazzled eyes sat Sara-Kate Connolly in a black rocking chair holding a long-legged figure in her arms. The figure's feet were dragging on the floor as Sara-Kate rocked back and forth. The figure's face was sad and white. Its hands clung to one of Sara-Kate's hands. Oh those elves! They were so ingenious. They had changed their own roly-knocking noises (the sounds of tiny feet, of mysterious wheels) into the noise a rocking chair makes against a bare wood floor. They had changed themselves into Sara-Kate and—who was it?—her mother? Hillary's hands flew up to cover her mouth. It was all so ridiculous!

At the same time, the scene in the room came to life. Sara-Kate's bullet eyes zipped across the floor. They shot into Hillary with a force that made her gasp. This was no trick. Sara-Kate had no sooner seen Hillary than she began to struggle to her feet. She lifted the thin figure from her lap and laid it in the chair. She whirled and rushed across the room.

"Out!" Sara-Kate screamed. "Why are you here? Get out!"

She leapt at Hillary like a wild animal, as if she meant to tear her to pieces.

"Sara-Kate! It's me!" Hillary tried to say, but her voice was strangled by surprise.

"Get away! Go back home!" Sara-Kate screamed. She grabbed Hillary's coat with both hands and pushed her out the door. Then she dragged her along the hall and tried to shove her down the stairs.

"Stop it!" Hillary cried. "It's me. It's me."

Sara-Kate did stop. She stopped long enough to pull Hillary's face up close to hers and to hiss like a furious snake.

"You get out and don't come back," Sara-Kate hissed in this new, horrible voice. "Forget you ever came here. Erase it from your mind. It didn't happen. You were never in this house."

Hillary stared at her in horror. She turned and began to run down the stairs.

"If you come back, you can bet you'll never go home again," Sara-Kate yelled behind her. "And if you tell anybody anything, even one little thing, that'll be the end. The end of you, I mean. The awful end. The final end. If you dare say one word to anyone on this earth, I'll . . ."

Hillary didn't wait to hear the details of what Sara-Kate was going to do to her. She was on her way out the back door at last. And she was crossing the Connollys' yard in three strides, and she was crashing through the hedge, back into her own safe garden.

Eight

There were times during the next week when Hillary thought she must have dreamed her visit to Sara-Kate's house. Certainly it was dream-like enough—a shadowy staircase, a secret room, a sense of unreality, of having been in a fantasy. For dreams, like fantasies, take place only in your head. Whatever happens in them, they stay in your head and leave few signs of themselves in real life.

So it was with this frightening visit. Though Hillary seemed to have been inside Sara-Kate's house, and though certain shocking events appeared to have happened, now her life went on in the most ordinary way, made up of the most ordinary things. Day after day, she did her homework, shopped at the supermarket with her mother, brushed her teeth, and combed her hair. Night after night, she watched television, was kissed good night by her parents, and fell asleep in her bed. Nothing was changed, and furthermore nothing was changed at Sara-Kate's house. It continued to look as it had always looked— gray, gloomy, in need of repair. At night, the windows were as dark as ever.

At school, people said that Sara-Kate had gone away with her mother on a trip. With relief, Hillary believed it must be true. There was no sign of her anywhere. No one came out of her house and no one went in. No one came to look after the elf village, or to fix the Ferris wheel, which had blown off the cinder blocks during a rain storm and lay on its side in the dirt. (Hillary had peeked through the hedge and seen it.)

No one talked about Sara-Kate, either. Now that she was not on hand to scandalize people with her gas-station boots, or to yell, or to sit by herself eating mush at lunchtime, there was no reason to discuss her. Sara-Kate Connolly was gone. Hillary had had a dream. The school office had received a note from Mrs. Connolly withdrawing her daughter from school, someone said. There was no need to look into the matter more deeply.

But when Hillary really thought about it, she knew she hadn't dreamed the empty house. She knew she'd been awake when she climbed the dark stairs. She knew the second-floor room had been real.

Hillary didn't want to think about these things. Hadn't she been told to forget?

"You were never in this house!" Sara-Kate had screamed. Now Hillary tried to make it so. She closed certain doors in her mind. She turned certain locks. She shut off the lights and walked away. But the memory of Sara-Kate holding her mother in the rocking chair in the upstairs bedroom would not be locked up. It followed her around like a determined dog.

"Go away! Go away!" Hillary whispered to this dog of a memory.

"What?" Jane Webster would ask.

"Did you say something?" Alison Mancini would demand.

She was back with her friends, and happy to be back. They were wearing their star jackets in school again, and when they walked down the hall, it was always three abreast and shoulders touching. They were the Three Musketeers, they said. Alison's mother had persuaded her to cut her hair off short as a boy's around the ears, and now Alison was trying to persuade the other Musketeers to have it done.

"You just walk in and ask for an Eton cut," she said. "They know how to do them there."

"Where?" Jane inquired.

"At the place my mother goes. They do everything, nails, eyebrows, skin. My mother had a facial the last time she went. They put these layers of cream on your face and massage it around and then wrap everything up in a hot towel. It makes your skin come out really soft and nice, like a brand new skin. The old, ugly skin just peels right off."

Alison glanced at Hillary. "You're whispering to yourself again," she informed her coldly, "and it's driving me crazy."

"What is it?" Jane asked with a concerned look. "You can tell us. We can tell each other everything. We're supposed to. tell each other, in fact. That's part of being a Musketeer. Listen, Alison, that gives me an idea. How about writing down some rules for ourselves, and then we can sign them and swear to obey, and then . . ."

"Go away. Go away," Hillary told the memory of Sara-Kate under her breath.

Whenever she thought of Sara-Kate in the second-floor room, Hillary thought immediately of the elves. She had expected the elves to be there. More than that, she had known they were there. She had felt their presence in the house as strongly as she felt the presence of her own mother when she was out of sight in another part of the Lenoxes' house. Hillary might not be able to see her mother but she could tell when she was nearby, working at some job, humming, making little tapping and rustling noises that were distinctly hers.

"Do you think I could have gone up those stairs without knowing what was up there?" she would have protested to Jane and Alison if the subject had been one she could talk about.

"Do you think I could have sneaked down that dark hall? The elves were there. They were in that room. They were there and then . . ."

Here Hillary came to an impasse. The elves had vanished and Sara-Kate and her mother had appeared. How could such a thing happen?

Evidence can have several different meanings, she remembered Sara-Kate teaching her. It can add up to different answers depending on how it is looked at. And that seemed the only way the problem could be resolved. For when Hillary added up the evidence on one side, it came to one unmistakable answer: elves! But when she looked at the facts from another point of view, there was no possible explanation but that Sara-Kate and her mother had been in that room all the time. Could both views somehow be true at once?

Whatever the case, neither Sara-Kate nor the elves had returned to repair the village. Hillary crept back to look again, under cover of evening. The little houses were more broken than before. The big house and the yard were silent, abandoned. She looked up to the window on the second floor and there, for a terrifying moment, she thought she saw something. A shape darker than the dim space of the window materialized before her eyes. Then it dissolved, became a trick of the mind. Sara-Kate was gone. The elves were gone. There was nothing, nothing in that old empty house.

"Go away. Go away," Hillary whispered. She flattened her hands against her ears and held them there.

"What?" her mother said.

"Were you talking to me?" her father asked.

"Hillary! What is wrong with you!" shrieked Jane and Alison. "If you have something to tell us, then tell us!"

Another week went by. A light snow fell. Ice formed on the town pond. In the cellar of the Lenoxes' house, Mr. Lenox started a building project.

"I'm making a trellis," he said to Hillary one night, after she had come down the gritty cellar stairs to see what all the pounding was about. "It's for the garden. I'm going to plant a trumpet vine near the house next spring and it will need something to climb on. It will need something to throw its long green arms around and pull itself up on inch by inch toward the sun until it has filled every space with leaves and produced the most brilliant orange trumpet flowers you've ever seen! Oh, yes, trumpet flowers!" Mr. Lenox crowed, while Hillary looked at him in alarm.

"I've been dreaming about trumpet vines at night," he added sheepishly. "Last night I dreamed I *was* one." He glanced at Hillary. "Do you miss your elf garden as much as I miss my garden?" he asked.

Hillary shrugged. "It wasn't really a garden," she said. "The only flowers were weed flowers, and there was a lot of junk lying around, but . . ."

Her father nodded. "I guess everybody has a private idea of what makes a good garden," he said. "Now, for me, the Connollys' backyard doesn't amount to much, and your mother thinks Sara-Kate isn't the best of all possible friends you could have . . ."

"But she is!" Hillary exclaimed with a sudden burst of warmth. "She is the best possible. I know it's hard to see, but Sara-Kate is a wonderful person. She's taught me all kinds of things. And she's talented, though she doesn't like to show it. Do you know she can walk on her hands? She walked up and down her driveway one time, and even up the steps of her porch. I couldn't believe it. But she never would do it again. She gets mad a lot if you say the wrong thing and then . . ."

Hillary stopped and glanced suspiciously at her father. It was the first time she'd told anyone about her feelings for Sara-Kate and now, having told this much, she felt a terrible temptation to continue. A flock of words was rising inside her. A hundred small details about Sara-Kate's habits and their work together, about the long, cool afternoons at the elf village sprang into her mind and she wanted to tell them.

But on the heels of the hundred details came the hundred questions. They were the questions that Sara-Kate had refused to answer and the ones that Hillary had learned not to ask. They were the questions she must never ask her parents or her friends, because to ask would be to tell and to tell about Sara-Kate was unthinkable.

Hillary pressed her lips together and stopped talking. She looked down at her father, who was on his knees working over the trellis. She saw Sara-Kate come charging at her out of the upstairs room, yelling, grabbing her, shoving her—almost down the stairs!—and her eyes filled with tears.

"You didn't need to do that," Hillary wanted to tell her. "I wouldn't have told. You could have said anything and I'd have understood. You didn't need to go away. You could have trusted me."

Mr. Lenox cleared his throat and half-turned toward his daughter.

"I saw Sara-Kate the other night, out late, coming home from somewhere," he said. "Running home, I should say. I almost drove into her crossing the Valley Road intersection."

Hillary stared at her father. "You saw Sara-Kate?"

"Probably out on another errand for her mother. It's a shame how she orders that child around."

"It couldn't have been Sara-Kate you saw," Hillary said. "She's not here. She's on a trip with her mother. Or maybe she's living somewhere else by now. They went away more than two weeks ago."

"It was Sara-Kate all right. She ran directly in front of my headlights. I jammed on the brakes, but she got out of the way in time."

Hillary watched her father, who was crouched over, pounding a nail.

"When was this?"

"A couple of nights ago." He took another nail from his pocket and hammered it in. The structure on the floor quivered with each blow.

"Wait a minute, I know exactly when it was," Mr. Lenox said, sitting up. "It was three days ago, last Monday about eleven o'clock at night. I was coming home from that town council meeting. I guess she came back without telling you," he said to Hillary.

"But she couldn't have. She's not there."

"How do you know?" Mr. Lenox asked.

Hillary felt a stab of fright in her chest. Then a stab of longing.

"She's not at school," she told her father. "If she'd come back, she'd be at school, right?"

Mr. Lenox took out another nail, lined it up, and hammered it in.

"Well, all I know is what I saw," he said. He rose to his feet with a grunt, picked up the half-made trellis, and handed it to Hillary. "Hold this thing up straight so I can measure it," he said, "and I'll tell you how Sara-Kate went home that night. She turned down Congdon Street, cut across the Briggs's yard to Hoover Street, cut through the Smythes' yard on the corner of Hoover and Willow, and ran into her own yard and around back of the house. I saw her go. I had to go the same way in the car to get home and I kept seeing her ahead, running in the dark. It was impressive, I must say. I'd never seen a person that small run so fast."

Nine ✗

Hillary slipped out the back door of her house like a fugitive, her quilted jacket rolled up and clutched to her chest. Behind her, she heard her father start to hammer again in the cellar. She closed the door quickly, went down the porch steps, and walked to a place away from the house where the porch light did not shine on her. Here she stopped and put on the jacket, zipping it tight around her neck. She groped for her mittens in the pockets.

The night air was frigid. The temperature was going down to ten degrees, and more snow was on the way. Her parents had talked about it during dinner. The storm might come as early as tomorrow morning.

"Welcome to winter," her mother had said cheerfully. "Maybe they'll have to cancel school," she'd added, smiling at Hillary.

Her wooly hat was in her pocket but the mittens were missing. Hillary wondered if they had dropped on the closet floor when she'd gone to snatch her coat, ready to leap into the closet herself if her mother appeared. Mrs. Lenox was upstairs reading, but it was too risky to go back now. The clock in the kitchen had showed just past 9:00 p.m., which was Hillary's bedtime, though her parents sometimes forgot and allowed her to stay up later. One thing they would never allow, however, was an unexplained, late-night walk in the cold by herself.

Hillary's unprotected hands were already stinging. She pulled the sleeves of her jacket down over them and gathered the sleeve ends with her fingers to close the openings. Then she walked across the driveway and went downhill into the dark. As she moved, her eyes sought the black bulk of Sara-Kate's house, and when she'd found that, she leveled her gaze at the second floor, to the windows just under the line of the roof. And when those were picked out (for her eyes took several minutes to accustom themselves to the darkness), she looked straight at the one window, the window on the right overlooking the yard, and beamed all her powers of detection there. If Sara-Kate was inside, Hillary was going to find out this very night, this very hour, because no one—not even elves—can stay inside a house at night in total darkness. And if there is a light, even the faintest candlelight, even the smallest flashlight, it will inevitably show up to those looking in from outside.

Hillary knew how ingenious light can be at escaping, because of her own attempts to read under her covers or inside her closet after she'd been put to bed.

"But how did you know!" she would wail when her mother caught her in the act and took the little reading lamp away.

"There was a glow," her mother would say. Or, "I saw some light coming through the cracks."

Now Hillary trained her eyes on the window and looked for cracks. She ran her eyes across the whole expanse of Sara-Kate's house and watched for glows. She came to the hedge and squeezed through just far enough for a clear view.

The house was as gray and unrevealing as the face of a cliff. Windows pocked the dark surface at regular intervals, but there was no sense of depth behind them. They were like unimportant chinks in a block of stone. In fact, it was the weight of Sara-Kate's house that Hillary felt more than any other thing at that moment, as if the place really were made of rock so dense that it had tipped the land it stood on. Down, down, it plunged into the black trough of Sara-Kate's yard, while behind Hillary, her own house was lifted up, bright and light as a feather, toward the starry sky.

Hillary shook her head and sighed. Her father must have been wrong. The Connollys' house was as deserted as it had been these past two weeks. In a way, she was relieved. Now she could climb the slope of her own yard, slip back inside her own house, and go to bed without anyone ever guessing she had been away. And tomorrow, perhaps, the snow would come, enough to go sledding this time. Her mother would make hot chocolate, and her father would tinker with the snow blower, which was always breaking down just when it was needed.

There was a long-standing family joke about it. And who knew? she might invite Alison and Jane to go to a movie with her tomorrow afternoon, or to build an igloo.

A black figure came out of Sara-Kate's house and sat down on the doorstep.

It came so quickly and unobtrusively that Hillary felt no surprise. The door made a tiny sound and then the figure was seated, slim and shadowy, on the step. Hillary leaned forward and held her breath.

Sitting motionless as it was, the figure was all but invisible. If Hillary hadn't seen it move before, she could never have picked it out now against the house. Was it Sara-Kate? Hillary strained her eyes at the shadow. She thought she detected the shape of a head turned away from her. She thought she saw an arm. Or was it a leg? She could see pieces of this shadowy person but she couldn't put it together into a whole.

She whispered, "Sara-Kate?" but so timidly that the name hardly left her lips. The shadow didn't move. Was something really there? Had she imagined it?

Then the shadow moved. It stood up and sauntered out into the yard. It was small, and thin as wire, and it was not wearing a coat. A dry crunching sound came from under its feet, which seemed heavier and bulkier than the rest. With its hands in its pockets, the shadow ambled across the yard toward the elf village. It made a wispy noise as if expelling breath. It bent over briefly to look at something, then righted itself. It moved on toward the fallen Ferris wheel.

"Sara-Kate!"

Hillary stepped from the bushes as she said the name a second time. But once in the open, she stopped.

"Hillary?" The shadow turned with what seemed to be a hint of eagerness.

"Sara-Kate? I wasn't sure it was you."

"Of course it's me. Who else would be walking around in my yard in the dark?" Sara-Kate leaned over and picked up the Ferris wheel.

Hillary approached her warily. The figure in front of her looked like Sara-Kate and talked like Sara-Kate, but something made Hillary hang back.

"I thought you were gone," she said. "I thought you moved away. You were never at school. You were never here."

"I was gone," Sara-Kate said. "But now I'm back. For a little while, anyway." She regarded Hillary through the complicated wires of the Ferris wheel she was holding up for examination. "It's not broken," she said about the wheel. "It can work again. Maybe you should come over tomorrow and help me clean up this mess." She waved her hand around the yard, ending up with the battered elf village.

Hillary followed the arc of that wonderful sweep of hand with hungry eyes. She wanted to come more than anything. She wanted to fling her arms around Sara-Kate's thin shoulders and hug her. But still she was suspicious.

"It's supposed to snow tomorrow," she said. "I don't know if I can come." She looked Sara-Kate in the eye and added, "The elves are back, too, aren't they?"

"Yes," Sara-Kate said. Hillary glanced away. She didn't need to be told what she could already feel. All around her, the yard was starting up again. She heard a faint humming noise coming from the overturned washing machine. She heard an infinitesimal clicking in the dead grasses, a rustle among the bushes.

Sara-Kate had leaned over to lift the Ferris wheel back onto its two cinder blocks. She centered the great wheel upon the metal rod and straightened some wires that had bent under the impact of the fall. When they were fixed, she stepped away to admire her work from a distance.

"Watch!" Sara-Kate commanded. Her hand swept the air again. Directly overhead came the sharp cry of a bird. It seemed impossible on this wintry night, with the temperature steadily dropping and a storm on the way, but there it was.

And then, more impossible still, the Ferris wheel began to turn. Slowly, haltingly, as if pushed by invisible hands, it moved around, once, twice. It picked up speed and started a more methodical spin. Though there had seemed to be little light in Sara-Kate's dark yard, the wheel's spokes were illuminated. They flickered past Hillary's eyes, faster and faster, until the wires and spokes were spun together into a silvery tapestry, and the Popsicle-stick seats flew out like golden rockets from the rim.

Then silently, by degrees, the Ferris wheel slowed. The wires became visible again. The Popsicle sticks drew in. The spokes separated themselves, and the big wheel wound down, darkened, and finally stopped.

Up above, wind churned the leafless branches of the trees, then blew past. Hillary blinked.

"Now will you come tomorrow?" Sara-Kate demanded in her ear.

Hillary nodded. She couldn't take her eyes off the wheel.

"Was it the elves who made it spin?" she asked. "It was the elves, wasn't it? But, for a minute, it looked as if . . ."

She turned in wonder to the thin figure beside her.

"Sh-sh-sh," whispered Sara-Kate. She beamed her tiny eyes on Hillary. "It's better not to talk about it."

Ten ✗

How Hillary, in her excited state, got back inside her house, out of her coat, and upstairs to bed without her parents seeing, she hardly knew. She nearly ran into her father coming up the cellar stairs, muttering to himself. But she dodged into the kitchen and he passed on to the bathroom, which gave her time to race up the front stairway and into her room.

It was ten o'clock exactly and she had just slipped under the covers when her mother looked in sleepily to see if she was still awake.

"What an independent child you are," Mrs. Lenox said, coming over to give Hillary a hug. "What did you do all evening? I never heard a sound, and now you've even put yourself to bed. You won't need a mother at all by next year. I'd better start interviewing for a new position."

"Silly," Hillary said, smiling up at her. "I'll always need a mother." But she offered not a word of explanation, and after her mother had gone she lay awake thinking wild and dazzling thoughts that made her feel quite separate from her parents and their ordinary lives.

For Hillary had seen an elf that night. She was sure of it. To lie still in bed and think everything through only made it clearer. All those days of peering into bushes, all those afternoons imagining faces in the leaves seemed ridiculous now when the real thing had been walking around in plain view the whole time.

How stupid she had been to suppose that elves must have pointed feet and little caps. How idiotic to think they must always be tiny. These ideas were held by a world that knew nothing about elves, by people who had never really looked, who were afraid to look, maybe, Hillary thought, remembering how she had pushed Sara-Kate's appearance in the upstairs room from her mind because it seemed so strange and frightening. Not that seeing an elf was easy even when you did want to look. Hillary had been looking at Sara-Kate Connolly for two solid months and only tonight had she finally begun to see.

Sara-Kate had thick skin not because she was "like an elf" but because she was one. Sara-Kate wasn't miniature or green but she had the elf's thin body and the elfin quickness. ("I'd never seen a person that small run so fast," Hillary's father had said.)

Sara-Kate ate elf foods like berries and mint leaves. She hid herself inside the sagging folds of her old clothes in the same way the elves hid within the junk and disorder of the Connollys' backyard. And how had she come to know so much about elves in the first place except by knowing them from the inside, by being one?

The elves in Sara-Kate's yard had not come to

live there by chance, Hillary now saw. Sara-Kate hadn't simply found them one day outside her back door as she pretended. The elves were there because Sara-Kate was there. She was their leader and protector. She kept their small community safe from the outside world. When Sara-Kate went away, the elves went with her. And when the weather grew too cold for even the thickness of an elf, she brought the precious magic beings inside to live in her empty house—an elf house, it must be—with her strangely sick mother.

Hillary lay in her bed shivering with the force of these thoughts. It seemed that her mind had become ten times sharper, ten times brighter, and that it could go into dark places that had confounded it before. Such was the energy of her imagination, that she wondered if she were becoming a bit of an elf herself. Was it possible to become an elf by associating with one?

Hillary stayed awake for hours that night. When she slept at last, she entered dreams that were filled with magic and the impossible possibilities of things, dreams that, oddly enough, were not so different from what was happening to her in her real waking life at that moment.

Hillary woke the next morning to a world in silent frenzy outside her window. Armies of snowflakes swirled before her eyes. The round outline of her father's garden was already erased and the birdbath had collected an odd-looking drift on top. It rose in the basin like a lop-sided white flame, giving the birdbath the unexpected look of an Olympic torch.

"A foot of snow fallen and another foot predicted," Mrs. Lenox informed Hillary when she arrived in the kitchen for breakfast. School was cancelled and "The snow blower's broken, of course," her mother said.

"Of course," Hillary replied.

"See that white mound crawling on its knees out there on what used to be our driveway?" her mother went on, gesturing out the window.

Hillary nodded.

"That's your father. He's dropped the screw-driver."

However, this snowstorm, like many of its relatives, had no intention of being cast in the role of predictable, and shortly after ten o'clock it tapered off to a sprinkle, then stopped. The sky cleared. The air warmed. Sara-Kate's house, which had been hidden all morning behind curtains of falling snow, came into view before Hillary's anxious eyes. She'd been half afraid the place would vanish during the storm, whisking Sara-Kate from her grasp again.

She was out the door tramping eagerly toward the Connollys' yard before her father had finished plowing the front walk. But then, seeing that Sara-Kate was not yet there, she hung back by the hedge. After all that had happened, she felt shy about entering without an invitation. The snow rose over her knees in places and had changed the appearance of everything. It lay in an unblemished white blanket over the yard, concealing all but the trees and the largest bushes, and giving the open spaces a virtuous, barren look.

The rusty washing machine had become a gentle rise and fall in this soft-rolling landscape. The piles of car parts, the tires, the glass, the rotten wood and tin cans were smoothed away. The house itself looked more respectable surrounded by such tidiness and dressed in snow garlands along its gutters and windowsills. And finally, as if these gifts of cleanliness and order were not enough, the sun came out suddenly from behind the last snow cloud and hurled a dazzling light upon it all.

Hillary stepped back into the shade of the hedge and hooded her eyes with one hand. She was not impressed by the snow's transforming powers. Where, she wondered, was the elf village? Had it suffocated under all this heavy beauty?

While the yard shone with the brilliance of diamonds, Hillary's thoughts plunged like moles under the snow to the dirty, junky places she knew and trusted. And she had just about figured out where the Ferris wheel stood, invisible though it was, and the approximate location of the little houses, when Sara-Kate emerged and issued the invitation she'd been waiting for.

"Why are you standing there staring like an idiot?" Sara-Kate yelled in a most irritating and un-elf-like voice. "Come on. Let's get started!"

These words set what was to be the disconcertingly ordinary tone of the morning, for not once did Sara-Kate reveal a flicker of elf-ness. Though Hillary longed for another sign, though she dropped hints about "elf magic" and finally asked Sara-Kate point blank if the Ferris wheel would spin again, the thin girl did not respond. She pretended to have forgotten everything about the night before. It was a great disappointment until Hillary reflected how "elf-like" even this behavior was. How could Sara-Kate be expected to cast her invisibleness aside all at once? Naturally she would find it safer to appear and disappear like her smaller relatives, to show only parts of herself until Hillary had proven trustworthy.

"Which I will," Hillary murmured with determination. "I will."

After this, Hillary stopped looking for signs. And indeed, Sara-Kate continued to play her role so convincingly that the whole issue began to seem rather silly in the light of the day, so snowy and free from school. And there was so much to be done! The village had literally to be excavated, house by house, stone by stone, like the ruins of Pompeii. Everything was there somewhere, but where? And how were they to find it without stepping on it first?

They divided the area into four sections and worked each section carefully and thoroughly in turn. Once a house was discovered, the gentlest fingers were needed to free it from the snow. This was slow work. Even Sara-Kate's hands turned numb and achy and had to be thawed out with warm breath, and then held in her pockets for a while.

After a house was unearthed, its yard could be dug out more quickly with mittened hands. Sara-Kate borrowed one of Hillary's mittens. But the stones in the little stone walls were always getting in the way and being knocked around.

"Let's just put them in a pile for now. Then we can lay them down in the right places when the whole village is cleared out," Sara-Kate suggested. "Also, all these leaf roofs have fallen apart and I was thinking that the elves might like wooden ones instead. There's a pile of wooden shingles under the back steps. Shall I get them out?"

Hillary nodded. She was working on a different problem.

"According to our calculations, the water well should be right about here," she said, pointing to a patch of snow she had been probing with a stick. "But, it's not. What could've happened?"

They found out a moment later when Sara-Kate stepped back from the house she had been working on. A muffled crunch came from under her boot.

"Oh, no!"

"It's a house!" cried Hillary, rushing over to look.

"But how could it be? There aren't supposed to be any here."

"And here's another!" exclaimed Hillary, just saving herself from putting her own foot on it.

Sara-Kate looked thoroughly alarmed.

"Wait a minute!" she said angrily. "Has somebody been building more elf houses in this yard while I wasn't here?" She gazed at Hillary, who shook her head.

"Then how could . . ."

"I know what it is," Hillary said. "We've figured the village out wrong, that's all. Look, the rest of the houses lie under the snow in this direction, not up there where we were looking for the well. And that means the well must really be just . . . about . . . here." She probed a patch of snow and nodded at Sara-Kate.

"It's here," she confirmed.

Sara-Kate seemed relieved.

"Whew!" she said. "I thought maybe these houses were multiplying by themselves during the night."

"Well, I suppose there's nothing to keep an elf from building more houses if she needs them, is there?" Hillary couldn't help saying. She sent one more meaningful look in Sara-Kate's direction but the older girl took no notice. She put her head down and started excavating the house she had stepped on. For the next half hour, no one spoke as the laborious work continued.

At last, however, the village began to emerge again. On all sides, dramatic peaks of snow towered over the little houses as a result of snow-removal operations. The peaks gave the village the cozy look of a hamlet nestled in the foothills of the mountains, though what the serious-minded elves would think of this, Hillary was not sure. Certainly, they would have more difficulty coming and going over the snowy terrain. Would they provide themselves with cross-country skis?

Hillary smiled at this thought. She was about to ask Sara-Kate for her views on the matter when she noticed her standing rigidly beyond the village, her face turned toward her house. She was looking at the window on the second floor, Hillary saw. Its shade had been drawn up. Some commotion was underway up there, a silent flutter behind the glass.

Hillary stepped forward and caught sight of Sara-Kate's face. It was as tense as a knotted fist, wholly absorbed in the action above.

Hillary took another step forward.

"Is it your mother?" she asked softly.

"Yes."

"Is she still sick?"

"Yes." Sara-Kate stared up at the window. "She wants me to come in." She sounded tired.

"It's all right. Do you want me to go home?" Hillary asked her.

"I guess so."

"All right."

Sara-Kate sighed and turned to look at Hillary. There, in Hillary's face, she seemed to see something that interested her, something new and rather amazing if her expression told the truth.

Sara-Kate blinked. She folded her thin arms across her chest and examined the younger girl again.

"What is it?" Hillary said. She felt that she was

standing in a spotlight. "Do you want me to do something?"

Sara-Kate looked at her. "My mother has been worse lately and she likes to have me stay near her," she said. "Do you have any money?"

"I could get some," Hillary said.

Sara-Kate stared at her.

"Without telling anyone," Hillary added quickly.

"We are out of things," Sara-Kate told her. "My mother likes coffee and milk. And sugar. We need bread and some kind of fruit. She likes fruit."

A moment of silence rose between them. Hillary glanced up at the window over their heads, but she couldn't see anything. She looked back at Sara-Kate.

"What else?" she asked the small, tense figure before her.

"Whatever." Sara-Kate shrugged. "Anything. It doesn't matter."

"Should I go to the store?" Hillary asked.

"Yes."

"Should I go right now?"

"Yes," Sara-Kate said. "If you can get some money."

Without another word, Hillary turned and began to go home. She walked steadily, in a dignified way, until she reached the hedge. Once through it, though, out of sight of Sara-Kate, she started to run.

Eleven

Not even in her wildest dreams would Hillary have done the things she now did if Sara-Kate had not asked her. Never would she have thought of doing them or, after planning, have carried them out with such a cold, clear mind.

In the next hour she would lie to her mother, she would steal twice, she would walk alone down forbidden streets, and transact business in a grocery store with the composure of an adult.

"It was no trouble at all," she would tell Sara-Kate afterwards, handing over the bag of groceries in the Connollys' kitchen. It was almost the truth. Hillary had never been so proud to be trusted with a mission in her life.

"What did you tell your mother?"

"I didn't tell her anything. She was upstairs. I took a ten-dollar bill from her wallet on the counter. Then I called up to her and said you'd invited me for lunch."

Sara-Kate grinned. "That must have surprised her."

"It did."

She'd run out the kitchen door before her mother could protest. She'd gone through the hedge into Sara-Kate's yard, then around the house to the street in front. Most sidewalks weren't shoveled yet so she'd walked on the slushy side of the road. She knew the way from her trips with Sara-Kate and she wasn't afraid, not even when a car honked at her for being too far out from the curb.

In fact, she'd felt the opposite of fear: a slow-rising excitement. The day was so bright, the snow was so deep. There was a lawlessness in the air, a sense of regular rules not applying, of their being cancelled, like school. Cars nosed along the streets in a bumbling way, avoiding drifts and stranded vehicles. Children waded like penguins through gleaming white yards, or built snow forts, or sucked on porch icicles. Office workers who should have been at their desks hours ago shoveled their driveways lazily and talked to their neighbors. It was all so breathtaking, so free and easy, that Hillary wanted to kick up her heels and turn cartwheels in the street. But she kept herself on course, kept her face blank. She knew that she was more lawless than anything in that day and must not draw attention to herself.

"That will be $13.05 please."

"$13.05!" Hillary looked into the unsmiling face of the man behind the cash register. "But I only have ten dollars."

"Then you'll have to put something back."

"But, I can't! I promised I'd . . ."

The man sighed, rolled his eyes, and leaned toward her over the counter.

"I guess it'll have to be the bologna," Hillary said quickly.

· 85

She hated to give it up, though. The bologna was Hillary's idea of something extra that Mrs. Connolly might like. Meat was good for you. It made you strong. She'd retraced her steps to the cold-cuts case to put the package back but in the end the place she put it was in her pocket.

"Lucky the pockets in this jacket are big," she said to Sara-Kate in the kitchen, though she still quaked inside to think of what she had done.

Sara-Kate glanced at her. "You have to be careful of the mirrors," she said. "They have mirrors high up in the corners that can show what you're doing."

It was her way of saying thank you, and Hillary answered with a nod. She knew they were speaking a special language now, and more than that, that she had passed a test and been ushered through a secret door. Next, Sara-Kate asked, rather formally because they were coming together so fast: "Please stay. I'll be right back."

She took the paper bag and went upstairs. The electric stove in the middle of the room had been turned on. Faint waves of heat came across the cold floor from the oven's mouth. Hillary walked over and sat in one of the armchairs in the strange room-within-a-room. Now she could see why it was arranged as it was. The old stove wasn't powerful enough to heat the whole room, but if you stayed near it, you could be warm. The fan on the stove was working. It blew the heat toward her in a soft, pleasant way, as it was intended to do, she guessed.

Hillary took off her boots. They were wet inside from her hike through the snow. She sat back in the chair. She leaned her head against the chair's padded interior and thought how exciting it was to be here, on this most unusual island in the midst of the everyday world. All around the Connollys' house, the town honked and bumped, clanked and thudded, without an inkling of the secrets held within. It gave Hillary a delicious feeling to be sitting in such a private place, to have come through the ordinary face of things into Sara-Kate's hidden world. She stretched her hands out toward the oven's warmth and waited for her friend to come back.

In a little while, she heard the tread of boots on the stairs. Then Sara-Kate appeared, still carrying the paper bag.

"Is your mother all right?" Hillary asked, jumping up.

Sara-Kate shrugged. "She's okay. She says you really can stay for lunch, if you want to that is." Her eyes skirted Hillary's.

"Want to? I'd *love* to!" Hillary said. "Do you know that you've never invited me in before? Not once. I mean the last time I was here it wasn't really . . . well, I just came by accident. I wasn't spying on you, honest," she ended quickly. Sara-Kate had given her a look.

"It's all right," the older girl replied. "I know you didn't tell."

"I wouldn't have even gone upstairs except that I

thought the elves were there," Hillary explained.

"Elves in this house?" Sara-Kate produced an explosive hoot.

"Well, I was sure they were, and I still think—"

"Hey!" Sara-Kate cut in. "We're wasting time. Let's have a party. Come on! We've got everything." Then, in one of her wild leaps of mood, she began to race around the inner room, snatching a knife from a drawer slot, tossing the bag in the air, dumping its contents on the stove top.

"Bologna sandwiches!" cried Sara-Kate. She trumpeted through her fists. "Toodle-tee-toot-tee-too. Charge! That's what they do at the football games at the high school," she said to Hillary. "I go over and watch on Saturdays when I feel like it. It's neat. Do you want to come next time?"

"The high school is way across town!" Hillary protested.

Sara-Kate didn't hear. She was blowing more blasts on her trumpet and charging into the making of the sandwiches. Two slices of bread down flat, slap, then a thick stack of bologna slices on top of each one, slap, slap, and more bread on top of that, slap, no mayonnaise, no lettuce, no mustard. So what?

"I'd like to go see a football game," Hillary had to admit in the middle of the slappings and trumpetings. "In fact I guess I'd love to."

They sat side by side on the ragged, falling-apart chairs. Sara-Kate devoured her sandwich like a lion.

Hillary took polite bites and chewed thoroughly, as she'd been taught at home.

"Want another one?" Sara-Kate was up and flying again before Hillary had finished her third bite. She grabbed two mugs from the shelves in the drawerless bureau and filled them to the very top with milk.

"Watch out!" cried Hillary. "They're spilling."

Sara-Kate giggled. "Would you like to see how Pierre the Package drinks out of a cup?"

"Who's he?"

"He's awful," Sara-Kate whispered. She rolled her tiny eyes. "He's horrible. I read about him in one of those newspapers they sell in the bus station. See, he's got no arms or legs. Just a little stump for a body. It's all wrapped up in cloth, like a package. Anyway, here he is drinking."

She lowered her lips to the rim of a cup and, with her arms bent at a painful-looking angle behind her back, slurped at the milk.

"Want me to do your cup, too?" she asked, looking around.

"Ugh. Okay."

"Pierre the Package had to learn to do everything with his mouth," Sara-Kate went on, more seriously, when she had finished slurping Hillary's milk.

"Don't tell me," said Hillary. "I don't want to know."

"He types letters to people by holding a stick in his mouth to hit the typewriter keys. He turns on lamps and faucets with his teeth. He makes sandwiches and feeds his dog. Yup, he has a dog, a cute little terrier that jumps up in his wheelchair and licks the mess off his face after dinner. And listen to this. The way Pierre reads a book is by flicking the pages over with his tongue."

"Ugh! Ugh!" Hillary covered her own mouth with her hands, a thing Pierre the Package wouldn't be able to do no matter how disgusted he felt, she thought suddenly.

"We should try reading that way sometime," Sara-Kate was saying. "Who knows when it might come in handy. You know, you can learn to do practically anything if you really want to hard enough."

"Sure, let's try it," Hillary murmured, while Sara-Kate tore into another bologna sandwich and poured herself another mug of milk.

"This is the greatest," Sara-Kate said, leaning back with the mug in one hand. "Isn't this the greatest party? Are you having a good time? See, it's not as bad in this house as you probably thought it would be."

"Bad?" Hillary said.

"I mean, I do okay here as long as they don't switch off the electricity. I try to keep that bill paid up. I used to get heat from a furnace like everybody else, but it broke. It takes big bucks to fix something like that. Upstairs I've got three good electric heaters. Usually I move us up there in the worst weather. Nothing much works down here when it gets really cold."

Hillary stared at the lean planes of Sara-Kate's

face and noticed for the first time the odd little dark marks under her eyes, like tired smudges in a grown-up face.

"One good thing is I've got two of these hot plates so I don't have to haul them up and down stairs," the elf-girl went on. "Wherever I happen to be when I want to cook, I can cook." She smiled. "I bet you don't have that in your house."

"No, we don't," Hillary said.

"And you don't have a big, friendly stove like this that you can lie back with and put your feet up on."

"No."

Sara-Kate stuck her work boots up on the stove and stretched out luxuriously in her chair. "And you can't have parties like this either, with just two people or whoever you want to invite, and whatever you want to eat. Are you still hungry? How about a cup of coffee? Whenever you've got a space left you can always fill it up with coffee."

Hillary leaned toward her suddenly. She put her hand on Sara-Kate's arm and said: "You do everything around here, don't you? You run this whole house."

Sara-Kate sat up. "What do you mean?" she asked, on guard in an instant.

"You keep pretending that your mother is the one telling you what to do, like everyone else's mother. But that's not right, is it? She doesn't tell you anything. She's too sick. You're the one taking care of her."

"That's not true!" Sara-Kate replied. "My mother

does tell me what to do. She tells me things all the time."

"But you're the one who does everything in the end," Hillary went on. "You buy all the food and do all the cooking."

"So what?"

"You pay the bills and wash the clothes and when something breaks, like the furnace, you decide what to do about it."

"So what?" Sara-Kate spat at her. "I learned how. I can do it. I help my mother, that's all. I bet even you have to help your stupid mother sometimes."

Hillary didn't get angry. She looked at Sara-Kate hard, as if she were trying to bring her into focus. "It's all right," she said. "I would never tell anybody. I was just imagining how it would be. What happens if there's something only your mother can do, like sign something, or talk on the phone? What if she needs to go somewhere, to the doctor or the hair-dresser?"

At that, Sara-Kate sagged in her chair. She sighed. She looked at Hillary as if she were two years old instead of nine, and folded her arms across her chest in that all-knowing, impatient attitude so character-istic of her.

"Look, whatever happens, I fix it," she told Hil-lary. "I sign it if it needs to be signed. I write it if it needs to be written. I learned my mother's writing. I talk on the phone, too, when it's working. I tell people what to do and they do it. Or, if they don't, I find some other way. I'm good at things like that.

My mother used to get upset all the time. Her mind's not always right, so then she gets sick. See, sometimes the envelope comes and sometimes it doesn't come. I learned what to do when it doesn't come."

"What envelope?" Hillary asked.

"You know, with the check, the money," Sara-Kate answered. "A lot of times, my father can't send it. He's not exactly rich. So then we run out."

"Run out! But then what do you do?" Hillary said, appalled. "How do you buy things like food and . . ." Her hands were rising up to her mouth again. She was looking at Sara-Kate over the top of them. "Like food and . . ." She couldn't think, suddenly, of all the things that Sara-Kate would need to buy. All the hundreds of things. "Like food and, you know," she ended lamely.

Sara-Kate shrugged. "I know," she said.

Twelve

The afternoon was passing. Through the torn shade of the window over the sink, Hillary detected the sun's shifted position. It was no longer overhead, hot and bright, but lower in the sky, half-screened by trees and neighboring houses. She thought that she ought to be going home soon. Her mother would begin to wonder where she was. She might be looking out the Lenoxes' dining-room window at this very minute, peering down at Sara-Kate's shadowy house: *Where is my child?*

Well, I'm here. Don't worry, Hillary answered her mother in her mind. I'm inside where I wanted to be, with Sara-Kate. And we're having a party in the magic inner room. At least I think it's a party.

Hillary gazed about herself and wondered suddenly if "magic" was quite the right word for this place, which now looked rather grim with the sun at its new angle. There was a hole in the floor near the sink, she noticed. Beside her, Sara-Kate was flipping her hair carelessly over her shoulder, preparing to answer the question about the envelope that didn't come, the money that ran out.

"So what *do* you do?" Hillary asked her again. She was met by yet another of Sara-Kate's weary shrugs.

"I get by. I know some ways."

"What ways?"

"People are always leaving their stuff around in a town like this. There was a whole shopping cart of food in the supermarket parking lot one time. At school there's lost and found. I could wear all designer clothes if I wanted. I don't take that kind of stuff, though. Who wants to look like those dumb show-offs?"

Hillary nodded.

"Hey, I should show you how to get into the movies for free sometime!" Sara-Kate exclaimed. "It's really easy. I'm not always broke, you know, but I never pay for the movies because it would be a waste. We should do it together. You'd see."

"What happens if you get caught?" Hillary asked uneasily.

"Who gets caught?" Sara-Kate's small eyes skimmed over her. "I bet you think I'm dumb because I got put back in school. That's what a lot of people think, and it's too bad for them. Just when they've decided how dumb I am and how smart they are, right then is when they happen to lose something. Something of theirs just disappears out the window."

Perhaps the sun had settled another inch. A finger of cold air caught the back of Hillary's neck and she shivered.

"You shouldn't do that," she said to Sara-Kate. "It's not right at all. You should ask someone for help instead of stealing all the time. If people knew you were living here taking care of your mother by yourself, they'd have to do something about it. They'd have to—"

"Wreck everything! That's right," Sara-Kate interrupted with a flash of anger. "Nobody knows how to take care of my mother except me. They've tried to do it. Even my father tried, but he couldn't so he left. Now I'm doing it. I've done it for a year so far and nobody even knows. People are stupid. They can't see a thing. They don't have a clue to what's going on right under their noses, in their own backyards."

Hillary stared at her.

"Do you know what would happen if I called somebody up on the phone and asked for help? Do you know what they would do?" Sara-Kate stood before Hillary with her hands on her hips and the whole rest of her body moving—twisting, jumping, quivering, kicking. It made Hillary think of the elf in her, the strange elf-ness that came at certain moments and then hid away again, came and went, so that Hillary could never finally decide who this small, fierce person was. She could never decide if she was cruel or warm-hearted, magic or ordinary, thick-skinned or fragile, a friend or a fraud.

"They would take my mother away," Sara-Kate said, without waiting for Hillary to decide this time either. Her voice had dropped to a whisper.

"But why?" Hillary asked. "Where would they take her?"

"They would put her someplace far away, out of sight."

"But why?"

"Listen Hillary, regular people don't like us, that's why. They don't like other people who live different from them, other people who are sick. They don't want us around. They don't want to look."

"Don't want to look!"

"So if you're thinking of going somewhere and getting help for us, don't do it. The only help we'd get is the kind that would look away and shake its head. Then it'd grab us by the neck and drag us off someplace we didn't want to go."

"But . . ."

"Help is the last thing you want to ask for when you're somebody like me," Sara-Kate told Hillary. "People like you can ask for help. People like me have to steal it."

Sara-Kate sat down abruptly in the other ragged chair. She sat without looking, knowing exactly, to the inch, where it was behind her. She knew everything about this crazy room-within-a-room because she had made it herself, Hillary understood. She was the one who had turned the drawers into tables, the bureau into shelves. She had positioned the fan on the stove to spread the heat around. She had organized everything, figured out everything, pushed everything together and forced it to work. And she had done it by herself. No one had told her, "Do it." No one had explained, "This is the only way." She was all by herself, separate from the world. She was her own single, strong, secret person.

Hillary moved toward her. "Are you an elf?" she asked Sara-Kate, who brought her head up with a jerk. "An elf," the younger girl repeated, longingly. "You can say it if you are. I'd never tell, I swear. I'll help you with your mother. I'll do anything you say."

Sara-Kate's little eyes had opened wider than Hillary had ever seen them. They looked surprised and puzzled, eager but undecided, and clearly Sara-Kate had something to say. She moistened her lips with her tongue. She pushed her hair off her forehead with the flat of one hand. She kept her eyes trustingly on Hillary and, in a minute no doubt, she would have spoken. She would have explained everything, allowed it to fall into Hillary's lap like a special present, the kind of present that is so precious one is tempted to keep it and not give it to anyone. But Sara-Kate was ready to give, Hillary knew. She was ready to tell at last and they were both leaning forward in their chairs, beginning to smile at each other, when the interruption came. It was a muffled shout:

"Hillary, are you here?"

And then again, from outside, while Hillary and Sara-Kate continued to stare at each other but with changing expressions: "Hillary! Hill . . . a . . . ry!"

"It's my mother. She's in the yard." Hillary jumped out of her chair.

98 ·

Sara-Kate leapt up, too. "Go and meet her, quick. Keep her away. She can't come in here."

"Where's my coat?" Hillary wailed. She looked around wildly.

"Hurry up! Get going!" Sara-Kate's hands had rolled themselves into fists. "She's coming to the door. I can hear her coming!"

Hillary found her coat on the floor between the chairs. She stuffed her feet into her boots and ran for the door.

"I'll come back later," was all she had time to whisper. Sara-Kate was waving her away frantically.

"Go on! Go on!"

"Hillary? Are you in there?" Mrs. Lenox's unmistakable voice came through the door, followed by the sound of knocking.

Thirteen ✗

It seemed so odd to be answering the knock of her own mother, to be opening a door that neither had met through before, that Hillary hardly knew what to say when Mrs. Lenox's familiar face appeared across the threshold.

"Good heavens! Here you are," her mother exclaimed. "I've been calling and calling. You've been here for hours. I really do think it's time to come home."

She gazed anxiously at Hillary, then looked past her into the room beyond.

"Goodbye and thank you!" Hillary cried loudly, on cue. She gave the door a yank to pull it closed, but it struck the side of her boot and bounced open wider than ever.

"Good heavens!" Mrs. Lenox said again, looking through the opening. "Is that you, Sara-Kate?"

Sara-Kate said nothing. She stood as if frozen against the side of the white stove in the inner room.

"What's going on here?" Mrs. Lenox asked, in a more determined tone. "What's happened to this room?"

"It's nothing!" Hillary cried. "It's just Sara-Kate. Come on, let's go!"

"Wait a minute," Mrs. Lenox said, with an om-

100 ·

inous note of concern in her voice. She stepped around Hillary and through the door.

"This place looks like some kind of fort," she said, addressing Sara-Kate. "What have you been doing? Where is your mother?"

There was a moment's pause, just long enough for a single, swift intake of breath. Then Sara-Kate moved forward with a practiced, gliding motion. Her tiny eyes zeroed in on Hillary's mother and her face composed itself into a mask of perfect politeness, an expression that Hillary had never seen on it before. Sara-Kate met Mrs. Lenox halfway across the room. She shook her hand in a most courteous and charming way.

"Hello, Mrs. Lenox. I'm so glad to see you again. I guess it's been a while. My mother is fine, but she's upstairs having a nap. I know this room looks terrible. We're having it fixed up so we had to move everything around. I'm sorry you had to come looking for Hillary."

"I did try to telephone," Hillary's mother put in.

"Well, the phone's been off since this morning, as you probably found out," Sara-Kate said smoothly. Hillary stood to one side, marveling at the ease with which Sara-Kate was inventing.

"It's the snowstorm, I guess," she went on. "There's a man who's coming soon to fix it."

Mrs. Lenox looked around helplessly. Something was wrong, but she could not put her finger on what it was that so alarmed her about the house.

· 101

RETA E. KING LIBRARY
CHADRON STATE COLLEGE
CHADRON, NE 69337

"Is the heat off, too?" she asked. "This room is so cold."

"They had to turn it off, just for an hour or two, so they could work on some pipes," Sara-Kate explained.

" 'They'?"

"You know, the workmen who are fixing up the house," said the thin elf-girl with the ragged wheaten hair. There was just a hint of irritation in her polite voice to let Mrs. Lenox know that she was imposing, that she would do well, now, to stop and go home. Hillary knew that Sara-Kate had put the irritation in on purpose, to trick her mother. It was an ingenious performance.

Nevertheless, Mrs. Lenox had scented something poisonous in the Connollys' house. Perhaps Sara-Kate was a bit too thin. Perhaps her hair was a trifle too ragged. Perhaps it was her boots, after all, that were her undoing. They looked so black and so shabby laced up on the end of her twig-like legs, as if she had stolen them off some drunken bum in the park, which perhaps she had.

"I believe I would like to speak to your mother anyway, if you don't mind," Mrs. Lenox said to Sara-Kate, cutting straight to the middle of things.

"I don't think she wants to be bothered," Sara-Kate replied evenly. "I'll have her give you a call when the telephone's fixed."

"No, thank you." Mrs. Lenox became more polite the more insistent she was. "Now, if you'll just tell her I'm here—or even better, show me where she is," Mrs. Lenox said, for she had seen a strange light flash in Sara-Kate's eyes.

"Please go away," said the thin girl. She stood directly in front of Hillary's mother.

"I don't know what you think you're up to," Mrs. Lenox said angrily, "but I am now going to speak to your mother."

"Mother!" Hillary cried. "Please don't! Please come!" She tried to pull her back by the arm, but how can a child pull an angry grown-up away from something she is determined to do? Mrs. Lenox walked forward despite the arms dragging her back, and the body blocking her.

Hillary let go when she saw how strong her mother was. She looked at Sara-Kate to see what she would do next. She hoped Sara-Kate had something up her sleeve, some further trick for escaping this predicament. If ever there was a need for magic, it was now. Mrs. Lenox walked across the Connollys' empty dining room, glancing around as though she had landed on an alien planet.

Hillary waited for the bird's sharp cry. She waited for a blast of wind, a streak of light, for an elf's miracle. Nothing happened. Her mother entered the front hall.

"What are you going to do?" Hillary whispered to Sara-Kate as her mother began to climb the stairs. Sara-Kate seemed not to hear. She pushed her hair out of her eyes with one hand.

"Sara-Kate! Do something! She's going to find out," Hillary cried when Mrs. Lenox turned left at the top of the staircase. They were standing side by side at the bottom, and Sara-Kate didn't answer. She stared toward the second floor.

Hillary heard the sound of a door opening up above. Then she heard her mother's voice:

"Good heavens! May I come in? I'm Helen Lenox from up the hill. Is everything all right?"

Sara-Kate turned her face slowly toward Hillary. She beamed her two tiny eyes straight into Hillary's eyes for one flash of a moment. Then she turned herself around and sat down on the bottom step.

"Don't be afraid," Hillary heard her mother saying gently in the second-floor room. "Don't be frightened. I'm here to help."

Fourteen

Sitting on that stair was how Hillary last saw Sara-Kate Connolly "in person" as she later thought of it. It was the last chance she had to say anything to Sara-Kate. But right then was when it was least possible to say anything, so there were no final questions, no good-byes.

Upstairs, Hillary's mother was speaking to Sara-Kate's mother in quiet, grown-up tones, and raising shades to let in the winter sun. Downstairs, Sara-Kate sat silent on the step, and even if one could believe that there were elves in the world, and that she once had been one, Hillary saw there was no magic in her now. There was no thick skin and no uncanny quickness. She was the same frightened child that any child would be whose family was in trouble. She was waiting on the step the way every child would wait to see what would happen next, what the grown-ups would decide.

Hillary's legs felt shaky, so she turned and sat beside Sara-Kate on the stairs, and they were so much the same size that their shoulders met exactly. Their bent knees rose to the same height. Their arms lay side by side in the same angles and attitudes. There Mrs. Lenox found them when she came back downstairs. She leaned over the girls and hugged them both at once. She told them everything would be all right now. She was going to get help.

Mrs. Lenox told Sara-Kate to stay with her mother until she came back with the help, and she told Hillary to come home with her, please. So Hillary got up and looked down at Sara-Kate.

"I'm sorry," she whispered, but Sara-Kate turned her face to the wall. Then Mrs. Lenox drew Hillary gently away and they walked toward the back door together.

Never again would Hillary see a house change as fast as the Connollys' house did in the days that followed. Almost from the hour of "the awful discovery," as people were soon describing the event, the place was transformed. Where it had been silent and empty, now it rang with noise: the clatter of feet, the chatter of voices, the snap of shades being raised. Where it had been dark, it was lit up like a stage. The house had been nearly invisible before, unnoticed in its drab and boring decay. Now the eyes of the neighborhood were upon it night and day and the smallest movements were cause for comment. Who was going in? What was coming out? Was that a light in the kitchen? Was that a squirrel on the roof?

Strange bits of information floated about. At least they were strange to Hillary, who, having lately been at the center of the Connollys' world now listened like a spy at the gossipy edges of groups. She listened to her mother speaking softly on the telephone.

Sara-Kate Connolly was in an orphanage and her mother was crazy.

Mrs. Connolly was in a hospital and Sara-Kate was crazy.

Sara-Kate's father could not be traced and relatives were being sought.

Then: Relatives had been found! They were coming.

Then: The relatives were here! From Michigan or Kansas or Montana. They were taking charge of everything.

"And thank goodness for that!" Mrs. Lenox exclaimed. "Now this town can stop worrying and get back to ordinary life. I've never seen such an array of prying eyes and nosy noses. Do you know that a newspaper reporter came by and tried to interview me this morning?" She put her arm around Hillary's shoulder as if to shield her from such attacks, but Hillary stepped away.

"What's going to happen to Sara-Kate?" she asked.

"I suppose she'll go to live with her relatives, poor thing."

"Poor thing! Sara-Kate isn't poor. She won't go anywhere she doesn't want to, I bet."

"She won't have much choice, I'm afraid," Mrs. Lenox said, smoothing her daughter's hair. "You mustn't worry anymore about either Sara-Kate or her mother. They are getting wonderful care and every sort of attention."

"Is it true that the house is going to be fixed up and sold?"

"You sound so angry! Of course it's true. The family needs the money and what else could they do with a house like that?"

"They could leave it alone and let Sara-Kate keep on taking care of her mother there."

"Hillary, that's ridiculous."

"She was doing okay by herself except for running out of money sometimes. And that wasn't her fault. That was because her father didn't send enough and she didn't dare ask anyone for help. She knew people didn't like her, that they wouldn't care. She was right, too," Hillary told her mother. "You should see what's happening down at her house. They're changing everything as fast as they can. They want to make it look as if Sara-Kate never lived there."

"I wish you'd stop going to the Connollys' house," Mrs. Lenox replied.

"I can't stop," Hillary said. "How can I stop? The elf village is there."

"Elf village! Hillary, after all that's happened I hope you still don't think . . ."

"No. I don't. I don't think anything except that Sara-Kate loved the village and she'd want me to take care of it while she's away," Hillary replied quickly.

She went every day to the Connollys' house and watched and listened to everything that happened.

A repairman came to fix the furnace. A telephone man came to reconnect the phone. The dead insects were swept off the dining-room floor by a gang of housecleaners who also made short work of the dust, the cobwebs, the grime in the bathrooms—none of which were working, they reported. It was a shame and a scandal. So the plumber was called to install new pipes, and the electrician was called to rewire the circuits, and a roof man came to fix a hole in the roof.

"Please be careful where you step," Hillary advised these workmen when they entered the yard. She crouched next to the Ferris wheel and placed an arm around it.

"See those little houses over there? See the pool? See the paths?" She pointed with her free hand. Yes, yes, they saw, grinning and winking like mischievous boys. The electrician cringed playfully:

"Is there something magic that lives here, then?"

"There might be. Who knows?" Hillary answered coldly.

"Should we be afraid of being changed into toads?"

"Only if you step on things," Hillary replied, glaring at him as she was sure Sara-Kate would have done. The plumber's work boots looked shockingly familiar.

✗ Hillary could stand guard over the elf village but she could not protect Sara-Kate from the things that people continued to say about her. She could not stop the whispered stories, the mean remarks. Even the newspaper, which Hillary had always thought of as an unbiased reporter of hard fact, came out with the oddest article.

"Is this story really about the Connollys?" she asked her mother. "Because Sara-Kate wasn't keeping her mother prisoner the way it says here. She was taking care of her when no one else could. And she was never 'dirty and dressed in rags.' Sara-Kate washed her clothes and her mother's clothes, too. She folded them carefully in the laundromat. I helped. She was always clean and the house didn't 'reek of garbage' either. It was just empty and dusty and strange. I was there. I saw everything."

Mrs. Lenox shook her head. "Are you sure," she asked Hillary, "that you were seeing everything clearly? Or were you just seeing what Sara-Kate told you to see?"

"No. No!" Hillary protested, but even as she spoke old doubts about Sara-Kate sifted into her mind and she found herself shouting to drive them back: "I know what I saw!"

"Well, whatever was going on down there I'm certainly glad I found you when I did," Mrs. Lenox said firmly, "or what would Sara-Kate have dragged you into next? Running her errands around town? Lying and stealing?"

Hillary flushed and lowered her eyes.

"We all feel sorry for Sara-Kate," her mother went on. "She's had a bad time. But have you ever wondered why she chose you for her friend? Why couldn't she have found someone her own age? Are you sure she really cared about you, Hillary? Or were you just someone who was useful to have around?"

"I think she cared about me *and* thought I was useful," Hillary said, with an angry upward glance, though suddenly she wasn't sure at all. She remembered how easily Sara-Kate had lied to her mother in the Connollys' kitchen. She thought of the older girl's secretiveness, her bursts of rage, her unexplained disappearances. Was it possible that she had not seen Sara-Kate clearly?

At school, the newspaper story fueled a new round of rumors and opinions:

Sara-Kate was a sad, misguided creature who'd been caught in circumstances beyond her understanding.

Sara-Kate was a sharp-eyed, street-wise kid who'd steal the coat off your back if you let her near you.

Sara-Kate was a nut. Hadn't she starved her mother half to death and refused to ask for help?

Sara-Kate was going to a reformatory. No, she wasn't. She was going to Kansas on a plane.

"To Kansas?" Hillary murmured in disbelief. She couldn't see what was true and what was not. There was no higher authority announcing, "This is the final truth!" The more Hillary heard about Sara-Kate, the farther away she went. Her small, thin figure was disappearing behind a screen of opinions and facts and newspaper stories, leaving Hillary in a place as dim as the rooms of the Connollys' house.

She was in a land of the unknown and the unknowable, she thought, a black land where not even her parents could help her.

"I am the only one who can decide about Sara-Kate," she whispered to the little elf houses in the Connollys' backyard. "Oh, if she would just come home."

The village comforted her. She kneeled in its midst, repairing roofs, straightening walls, while around her the wonderful yard that had sheltered the tiny community was invaded, laid open to strangers' eyes, littered with the workmen's debris. Its privacy and secrecy evaporated as Hillary watched, and was replaced by the cheapness and indifference of a run-down city park.

"If only Sara-Kate would come out her back door with her usual shout: 'Let's get going!' Then there would be no worry about what to do or what to believe. Then we could start all over again just being friends," Hillary whispered to the village that huddled like a real village at the feet of real snow mountains.

🦋 "Is it true that Sara-Kate left yesterday on a plane with her relatives?" Hillary asked her mother one afternoon, after school.

"I think it's true," Mrs. Lenox said, looking up from the potato she was peeling for dinner.

"She never came back to say goodbye. She never called."

"Well, I suppose there wasn't time in the end."

"I think she hates me because I'm the one who got her caught."

"No, no! Of course not! It wasn't your fault. It was no one's fault," Mrs. Lenox cried, running across the kitchen to hug her. But Hillary turned away with tears in her eyes.

Fifteen

In the dark winter days that followed, what Hillary missed most about Sara-Kate, oddly enough, were the very things that had made her so difficult to get along with: her sharp remarks and clear, cold eye. Beside Sara-Kate's crisp manner, beside her quickness and lightness, the girls at school seemed slow and heavy. In fat-faced groups they clumped through the halls, weighted down with fashion clothes and expensive book bags. They pouted and complained, gossiped and giggled, and Hillary watched as if she'd never seen such behavior before, as if she'd never belonged to such a group. She was outside all groups now, but not because she was excluded. Everyone was being rather nice to her, actually.

Her teachers asked after her health. People smiled at her in the halls. Jane and Alison were always putting their arms around her, guiding her toward private nooks where they could whisper together.

"Don't worry if you still feel a little bad about Sara-Kate," Jane said. "She was even worse than we thought, and she got you tied up in complete knots."

"My mother says it would take anyone a little while to get over something like this," Alison added, patting Hillary's hand. "It's not that you were stupid

114 ·

and fell for all the lies Sara-Kate told you, even though you did. It's that Sara-Kate was so terrible. Imagine going to the trouble of cooking up that elf village—"

"Which we finally went over and saw after she left," Jane interrupted. "It's no big deal as far as I'm concerned."

"Me either," Alison said. "And then imagine her making up that whole complicated world of elves, down to the tiniest details of what they like to do and what they like to eat."

"How did you find out about that?" Hillary asked angrily.

"Your mother told our mothers," Jane said. "Don't worry. We understand. We don't blame you at all. It wasn't fair to pick on someone so much younger. We blame Sara-Kate."

"Well, I blame you!" Hillary suddenly found herself yelling at them. "For not understanding one thing that happened. I blame you and I blame everybody in this whole dumb school!"

Not that she understood any better. She didn't. It was what made her so angry at them all, so angry at Sara-Kate, too, when she let herself admit it. If not for the fragile village, which day by day seemed more endangered by the yard, Hillary might have turned her back on everything. She might have walked away up the hill to her own house and shut the door for good.

Why had Sara-Kate left the village behind anyway? she thought crossly. If Sara-Kate cared for it so much, if it was really the magic place she'd pretended, why hadn't she taken it with her, or dismantled it and hidden it in some safer spot? Sara-Kate had gone without a word about the village, as if the place meant nothing and Hillary was nothing, too.

And yet, even as Hillary accused Sara-Kate, another way of looking came into her mind. She had only to approach the village for her bitter arguments to be grasped and whirled around, to be turned inside out by invisible forces. Couldn't it be argued, for instance, that Sara-Kate had left the village behind on purpose? Suppose she had left it as a present for Hillary, or as a sign of friendship. Perhaps it was meant to be a message of sorts, the very message that Hillary longed to receive: "Goodbye. I am all right. I'll stay in touch." To be around the village was certainly to find oneself, willy-nilly, in touch with Sara-Kate. She might no longer be there "in person," but she was there, Hillary discovered.

From her post beside the Ferris wheel, she watched the workmen come and go through the back door that only Sara-Kate had used before. Through the kitchen window, she saw a team of men move the stove back against the wall. How had Sara-Kate ever moved it in the first place? she wondered.

("Elves are strong. And magic," she heard Sara-Kate say in her ear.)

Hillary began to recognize certain real-estate brokers who came to direct house improvements, and she invited herself inside with them to look at the improvements close up: new tile on the kitchen floor; new counter tops; a door for the oven, and new burners. These things were certainly better than the shabby little room-within-a-room that was there before. And yet, there had been something wonderful about that other room, Hillary thought, something in the way the furniture had been taken apart and put together again so strangely. It was as if an entirely different sort of brain were at work behind it.

("Strange and little!" Hillary heard Sara-Kate's angry voice say again. "If you were an elf you wouldn't feel strange or little. You'd feel like a normal, healthy elf.")

A sale of house furniture and goods was announced and a women's group came to polish up the few remaining tables and chairs. Then people arrived to prowl and buy. Hillary prowled with them. She saw Sara-Kate's hot plates being sold, one to a bearded man with a limp, the other to a woman wearing bedroom slippers instead of shoes.

An oriental gentleman with a small and oddly elf-like figure bought the electric fan. He tested it first by lighting matches in its airstream. Then he picked the fan up and shook it like a stubborn catsup bottle. What on earth was he going to use it for? Hillary wondered.

("Why do you think these elves are anything like you?" she heard Sara-Kate ask. "Maybe they're so different that nothing they do is anything like what you do.")

Often on her visits inside the Connollys' house,

Hillary went upstairs and down the hall. She walked into the second-floor room where Sara-Kate and her mother had lived during the very cold weather, where they had hidden when Mrs. Connolly had grown too ill to be left alone and Sara-Kate had stayed with her.

It was empty now, of course. Everything had been taken downstairs to be sold. But those four blank walls still held a glimmer of enchantment for Hillary. She remembered how the door had seemed to bulge with light, how near she had felt to the elves' magic. There were other explanations for the magic, now. There are always other explanations for magic, Hillary thought.

"Sara-Kate was very smart," Mrs. Lenox had explained. "She knew that if she didn't come to school, the school would come looking for her. So she wrote a note withdrawing herself from classes. She used her mother's handwriting and her mother's signature and completely fooled everyone.

"Then, since she had said they were away on a trip, she was careful to keep them both hidden during the day. She didn't answer the door. She kept the shades drawn. The heat was off, of course, because the furnace was broken, and no one could tell they were there at all. At night, Sara-Kate came out under cover of dark while her mother slept. She went for supplies. She must have gone to different stores so as not to be recognized, and of course she didn't always pay for what she took. The night your father saw her she must have been coming home from one of these trips."

"Maybe," Hillary had answered softly. She'd been thinking about Sara-Kate's strange eating habits, about the "delicate stomach" that required hot cereal for school lunch but could suddenly take on large numbers of bologna sandwiches on special occasions. Had Sara-Kate eaten Cream of Wheat because it was the cheapest thing she could get that was hot and filled her up? Perhaps she really didn't like wild berries and mint at all. But then again, maybe she did. Hillary shrugged and glanced at her mother. Perhaps being hungry and cold and angry and alone didn't mean you couldn't still be an elf. In fact, maybe those were exactly the things elves always were, Hillary had thought, as she stood gazing up into her mother's face.

The village looked fragile, but it had staying power. From the window of the second-floor room, Hillary looked down on it, over the new wooden roofs, over the tidy front yards. The Connollys' house brought Sara-Kate back in stray bits and pieces, but the elf village was where she came back all together in Hillary's mind. More and more, the village seemed the only true thing about her, the only fact that was sure.

Here Hillary had first run into Sara-Kate's tiny eyes and felt the tiny eyes of elves upon her. Here she had watched Sara-Kate work coatless in the cold and learned about thick skins and private languages.

Hillary had only to crouch between the little houses to see Sara-Kate flick a strand of wheat-colored hair over her shoulder.

("It isn't where you look for elves so much as how you look," she would hear Sara-Kate say. "You can't just stomp around the place expecting to be shown things. Go slowly and quietly, and look deep.")

Look deep. Every day Hillary looked. If she had not yet seen an elf, if she still couldn't be sure of Sara-Kate, it must be because she was not looking deep enough, she decided. She redoubled her efforts, in the upstairs room, in the yard, on the streets of the town, in the whole world for that matter. There was no place safe from her watchfulness now, and no person either. She felt her eyes turning tiny, like Sara-Kate's. She felt herself turning shrewd.

Out in the Connollys' yard, she hovered protectively over the little well. Its bottlecap bucket was frozen in place, but come spring it would work again, she thought. The Ferris wheel had stayed upright on its metal rod. Every afternoon, Hillary walked to it and turned it with her mittened hand to make sure it still worked. It always did.

The elves' sunken pool looked more like a skating rink. Remembering the power rafts, Hillary leaned over and tried to see special marks of activity. Sometimes there were none, and a dark feeling would come upon her. But more often, strange scratchings appeared on the ice, or a mysterious circular clearing would show up in the snow nearby, and Hillary's heart would beat faster. She would glance toward the Ferris wheel and see again how it had glowed and spun on that extraordinary night, and hear the bird cry that had sounded when it seemed least possible. She would remember how Sara-Kate had trusted her and been betrayed, how she had revealed herself and been hurt, and how every single thing Sara-Kate had taught her about elves had turned out to be true about the thin girl herself. Then Hillary was sure that she had been in the presence of an elf, and that the village was a special, magic place.

"A place that's got to be saved," she told her mother one day, not long after another rumor had swept the street: a new family wanted to buy the Connolly house; a nice family with a dog and two children.

"Saved?" Mrs. Lenox asked with a frown.

"Moved," Hillary explained. "I'm going to move it into our yard. That way, when the Connollys' house is sold, the village will still be here in case anyone wants it again. A place like that shouldn't be allowed to fall apart. It needs to have someone taking care of it."

Mrs. Lenox shook her head in a despairing way.

"Well, I don't know. You'd have to put it somewhere out of sight and out of the way. Your father has the garden laid out so carefully. We wouldn't want the Ferris wheel sticking up in the middle, and those little huts would get caught in the mower if they were put on the lawn."

"Oh, no. They couldn't possibly go there," Hillary agreed. "How about behind the garage?"

"But that's not a place at all. It's full of rocks and briers."

Hillary nodded. "It'll be perfect," she said. "I was checking it over this afternoon. It looks so terrible that I guess I never thought of it before. It just goes to show."

"Goes to show what?" Mrs. Lenox asked, but Hillary had gone out the back door into the yard again, and there was no answer.

"It's getting rather dark out there, and cold!" Hillary's mother called to her, opening the storm door a crack so her voice would be heard. "I think you should come in now. Hillary! Where are you?"

Very odd, but there was still no answer, and Hillary seemed to have disappeared.